G000015747

BTexact COMMUNICATIONS TECHNOLOGY SERIES 4

Internet and wireless security

Other volumes in this series:

Volume 1 **Carrier-scale IP networks: designing and operating Internet networks** P. Willis (Editor)
Volume 2 **Future mobile networks: 3G and beyond** A. Clapton (Editor)
Volume 3 **Voice over IP: systems and solutions** R. Swale (Editor)

Internet and wireless security

Edited by
Robert Temple
and
John Regnault

The Institution of Electrical Engineers

Published by: The Institution of Electrical Engineers, London, United Kingdom

The Institution of Electrical Engineers,
Michael Faraday House,
Six Hills Way, Stevenage,
Herts. SG1 2AY, United Kingdom

British Library Cataloguing in Publication Data

Internet and wireless security
(BTexact communications technology series; no. 4)
1. Computer networks–Security measures 2. Wireless communication systems–Security measures
I. Temple, R. II. Regnault, J III. Institution of Electrical Engineers
005.8

ISBN 0 85296 197 9

Typeset by Mendez Ltd, Ipswich
Printed in the UK by T J International, Padstow, Cornwall

CONTENTS

PREFACE

As we move into the era of the next-generation Internet, user trust and confidence remains at the top of the agenda. Time and market surveys tell us that concerns about Internet security are impeding the widespread adoption of electronic commerce.

Underpinning next-generation networks and applications will be a range of security technologies to ensure that businesses and consumers can be confident in their use of electronic commerce and Internet technologies.

Of course, BT has a long heritage in the arena of electronic security stretching back to the pioneers from Dollis Hill that built Colossus. As you will discover from the content of this book, cryptography and computing are still very much at the heart of securing electronic communications.

Engineers from BTexact Technologies remain to this day at the forefront of the development and evolution of the security technologies that are key to the future of secure communications and the widespread proliferation of electronic commerce.

Chris Earnshaw

BT Group Engineering Director and Chief Technology Officer

INTRODUCTION

The chapters of this volume are grouped into a number of topical areas — demonstrating that the security canon is a very varied one. As Chris Earnshaw so rightly highlighted in his preface, getting the appropriate security built into new products, systems and services is fundamental to a rosy Internet future. We highlight a number of the vital issues to address as part of that process.

Chapter 1 reminds us of the importance of securing the infrastructure. Only when this has been done can the applications which run on it be deployed with any confidence from a security perspective.

The next two chapters introduce functionality which is starting to play an important part in the application security story. XML, the next-generation of HTML, is now viewed as the standard way information will be exchanged in environments that do not share common platforms. Special purpose XML languages and standards are practically announced daily. Chapter 2 describes XML digital signatures and XML encryption, while Chapter 3 explains how these can be used to provide Web-based security services.

An even newer idea is SecML, the Security Modelling Language. As systems become more complex, so the task of designing an appropriate security model becomes ever more challenging. Chapter 4 introduces a new approach which may help in this task.

XML will be an increasingly important technology underpinning public key infrastructures (PKIs). Chapter 5 tells us what else is new in this area. Chapter 6 shows us how both XML and PKI can be used in constructing an archival service for high-value data. This is one of a number of value-added services we expect to see deployed in the medium term.

It is now well-known that public key cyptography was invented by CESG several years prior to its parallel invention as Diffie-Hellman and RSA. A somewhat different approach to those models is an identifier-based public key cryptography. By a neat reversal, Shamir (the S in RSA) first proposed the concept but the first practical solution was invented by CESG. Chapter 7 gives an overview of that system.

We turn now to the wireless world where authentication and confidentiality are prerequisites for a commercial service. Chapter 8 leads us through the evolving 3G standards, and highlights the possibility that a non-PKI-based approach may prevail.

An alternative view, based on what has been happening in the WAP Forum, is presented in Chapter 9.

Another wireless service is TETRA, the digital trunked radio standard. An enhanced version is being deployed now in the UK, initially in the emergency services market. Chapter 10 explains how security is handled in this service.

Our next area is IP data networks and particularly virtual private networks. Chapter 11 explains how the adoption of the IPsec standard enables this to be done securely. Chapter 12 puts this into a real-world business context with its description of BT Ignite's virtual private data network service and its development.

Evolution towards integrated global networks makes all operators and users of their networks equally exposed to malicious attacks. Such attacks can be mounted almost anywhere in the world by individual crackers or even by an agency of a rogue state. Chapter 13 describes how BT's Information Assurance Programme deals with this enhanced threat.

Identification, authentication and access issues underpin all of these areas. Chapter 14 shows how biometrics or 'something you are' can help in conjunction with 'something you know' (a password or PIN) and 'something you have' (a token). BTexact Technologies and University College, London are collaborating in a study to determine how computer users manage their passwords now, and how to help users manage them more securely in the future. Many of the problems are well known to people in the security world — large numbers of passwords have to be memorised (average of 16), with users mostly writing them down, and lax attitudes to their security leading to insecure ways of working in general. Based on this research, Chapter 15 presents an alternative approach to improving access control.

Finally, Chapter 16 reminds us that to ensure good security we must manage the problem systematically. It describes how BS 7799 (ISO 17799), the Standard for Information Security Management, can be applied to achieve such a result.

In conclusion, we would like to thank all of the authors for their excellent contributions to this book. We believe that the end result will enable you to understand more fully what a great deal there is to worry about when you try to construct an appropriately secure network or system, and how you start going about addressing those concerns.

John Regnault
Security Solutions Manager, BTexact Technologies
john.regnault@bt.com

Robert Temple
Chief Security Architect, BTexact Technologies
robert.d.temple@bt.com

CONTRIBUTORS

C Awdry, IP Security Product Marketing, BT Ignite

C W Blanchard, Security Risk Analysis, BTexact Technologies, Adastral Park

K P Bosworth, PKI Solutions, BTexact Technologies, Adastral Park

S M Bouch, Principal Design Consultant, BT Affinitis

S Brostoff, Security Mechanisms Research, Department of Computer Science, UCL

C J Colwill, Information Assurance Consultant, BTexact Technologies, Cardiff

G P Fielder, Network and Information Assurance, BT Wholesale

D J Gooch, Network Security Solutions, BTexact Technologies, Adastral Park

J Hill, Network Security Solutions, BTexact Technologies, Adastral Park

M Hogg, Security Futures, BTexact Technologies, Adastral Park

S D Hubbard, IPsec Solutions, BTexact Technologies, Adastral Park

S A Jaweed, Security Research, BTexact Technologies, Adastral Park

M J Kenning, Information Assurance Services, BTexact Technologies, Adastral Park

I Levy, Identifier-based Public Key Cryptography Research, CESG

M W Moore, Network Security Services, BTexact Technologies, Adastral Park

C Natanson, Secure Business Services, BT Ignite

D W Parkinson, Security Consultant, BTexact Technologies, Adastral Park

M Rejman-Greene, Security Technologies Consultant, BTexact Technologies, Adastral Park

M A Sasse, Interaction Design, Department of Computer Science, UCL

A Selkirk, PKI Solutions, BTexact Technologies, Adastral Park

G Shorrock, Business Solutions Design, BTexact Technologies, Adastral Park

M R C Sims, Security Consultant, BTexact Technologies, Adastral Park

M F G Smeaton, formerly iTrust Security Infrastructure, BT Retail

N Tedeschi, PKI Solutions, BTexact Technologies, Adastral Park

M C Todd, Security Technologies, BTexact Technologies, Adastral Park

N T Trask, PKI Solutions, BTexact Technologies, Adastral Park

D Weirich, Security Mechanisms Research, Department of Computer Science, UCL

T Wright, PKI Solutions, BTexact Technologies, Adastral Park

1

BUILDING ON ROCK RATHER THAN SAND

M Hogg, S M Bouch and M F G Smeaton

1.1 Introduction to iTrust and the eBusiness Environment

The Internet protocol (IP) is open, offering ubiquitous access to BT systems and data for BT people and BT's customers. It is this very openness which presents tremendous opportunities for eCommerce. The downside, however, is that IP exposes BT to harmful attacks. The purpose of the iTrust Programme is to provide cost-effective protection to BT systems and data. It is a large programme involving the implementation of several software packages. There is a phased approach to implementation.

iTrust aims to minimise the damage and costs of attempted security breaches in an increasingly hostile environment; it has two main elements:

- replacement over time of the current disconnected security measures deployed for individual systems with an integrated strategic scheme — this would solve the growing problems of accessing multiple systems and would avoid escalating essential security administration costs;
- additional measures on the external firewall, and completion of a pilot system both to defend against internal hacking and to provide a second line of defence against external intrusion.

This chapter describes activity under way to support the former.

During the year 2000 the current BT firewall turned away three times the number of unauthorised entry attempts as it had in the previous year. Most of these are harmless but some are recognisably deliberate and serious enough to merit investigation by BT Security. New threats appear regularly.

BT invests substantially to secure data on its operational support systems (OSS). £1.5m was spent on CSS (customer service system — BT's main customer handling system) alone between 1994 and 1997. Such security, however, frustrates both BT people and their customers who need access to multiple systems, and causes integration problems when automating processes across these systems. The principal drivers for these investments include the following three areas.

- Legal and regulatory requirements

  These cover customer and employee data privacy and accuracy, non-disclosure of customer information and confidentiality of data in transit (Data Protection Act 1998 [1]). Regulatory obligations constrain information sharing with some BT subsidiaries. Stock Exchange regulations (Turnbull Report [2]) oblige companies to show that they are managing the business risk.

- Mandatory BT commercial security policies

  These enforce sound commercial protection of BT and customer IT assets to protect information integrity, prevent (and detect) disclosure of company information, maintain system and information availability, guard against virus infection, and prevent misuse of the Internet and e-mail. Key policies control the identification of users of BT systems, the auditing of their actions and all connections into BT's corporate data network.

- eBusiness commercial drivers

  These extend the need to identify users of BT systems and to log the actions of about 8 million self-serve customers. Integration and automation of common activities and administration is essential to avoid costs and simplify problems arising from distributed administration systems. There is a need to provide scalable single-user sign-on facilities for Internet-based services for employees, for customers, and for contracts over the Internet that cannot be repudiated. Business-to-employee, business-to-business (B2B) and business-to-consumer (B2C) transactions have slightly different issues and solutions are planned to be delivered in stages.

iTrust underpinned the transformation programmes, eBT, Options 2000 and the mobile workforce. It fully supports the approved Internal Audit and BT Corporate Security strategies for eBusiness, which certify the iTrust deliverables. Systems using iTrust components will have automatic certification for those components and overall system certification will be simpler. iTrust endeavours to employ standard industry techniques which may also be exploited by BT in its commercial dealings with customers.

So what options did BT have, set against the business environment:

- carry on with the current system-by-system approach to security;
- provide shareable, or reusable, bought-in components that will spread the cost across all users?

BT chose the second option.

1.2 Benefits of an Integrated Approach

Substantial savings in a number of areas are anticipated as a result of this integrated approach.

Security administration across a large number of discrete platforms is complex and results in substantial replication. For example, a user who needs access to 15 systems will require 15 accounts, 15 userIDs, 15 passwords, etc. Rationalisation of access control results in savings, through:

- reduction in current internal security administration costs through centralisation, automation and Web enablement for self-service;
- cost avoidance through productivity improvement for BT users from single sign-on (equivalent to 1 hour per full-time employee per year);
- savings by administration functions for password resets;
- reduction in development costs because the project utilises infrastructure components rather than develops its own solutions.

The main benefits from iTrust are in the costs that it will avoid as compared to a piecemeal approach to systems security. The single sign-on component will produce labour efficiency savings across BT, and, in addition, the current processes for administering systems security would increase by an estimated £24m per annum in response to the 8 million external customers given access to BT systems if no changes were made.

1.3 Architecture

The aim is to provide a set of services and functionality that can be used across the business. The issue of platform security is outside the scope of this work, and system owners will still have to consider the nature of the assets for which they are responsible and protect them. However, much of the burden of taking care of the security of the platform will be removed into the iTrust infrastructure. It is not necessarily the case that iTrust will provide one set of hardware serving both internal and external facing needs. In fact, there may well be duplicate hardware. However, the knowledge and experience gained within the business will result in economies of scale and sharing of experience. In addition, technical interoperation may be possible, providing enhanced end-to-end security.

To provide full security for an application, many aspects have to be taken into consideration, including:

- physical security;
- access to BT's network;
- network access to the application;

- platform security;
- manual processes and procedures;
- application security.

The architectural framework and implementation considers these factors.

The objective of the architecture is not necessarily to solve specific requirement solutions (although some may actually result from the design work), but to provide a set of coherent and integrated services which are available to applications, objects, elements and systems, in order to provide security functionality.

The architecture should be able to provide:

- automation;
- common mechanisms;
- end-to-end security;
- resilience;
- object model independence;
- design for evolution.

The current BT strategy is for the deployment of thin client computing throughout the business. It is known that there are some applications and circumstances where thin client computing is not feasible and therefore there will be a continued need for thick clients. It is intended that the architecture supports both.

Devising and implementing an architecture is not without its risks. During the design stages a number of technologies were new and unproven, for example:

- Windows 2000 Kerberos implementation was non-standard and the migration path for compatibility was not clear;
- IPsec was an immature technology and the means by which security associations were managed by a directory service were not clearly understood;
- directory information interchange standards were being formed;
- public key infrastrucures were relatively new in real-world deployments.

1.4 Architecture Overview

The architecture needs to provide the means by which it is possible to:

- uniquely identify each user;
- uniquely identify each object/application if required;
- determine a level of confidence in the application or client credentials presented and so determine trust;

- provide a range of security credentials so that, if required, an assurance can be provided that credentials have not been compromised;
- easily, and centrally, administer access control and authorisation policies.

Table 1.1 outlines the set of high-level requirements for iTrust, and Fig 1.1 shows the broad functionality that iTrust will deliver. An integrated administrative system underpins all the functionality within the domain, and will itself use the authentication and authorisation service.

Table 1.1 The set of requirements for iTrust.

Trust levels	The authentication and access control processes will need to support user and network environments that require anywhere from low to very high levels of security.
Single sign-on	The user explicitly authenticates to the authentication service and subsequent transactions are undertaken with reference to this.
Ubiquitous access	Users may authenticate themselves from varying access points such as internal network, remote access, and different buildings.
Multiple access modes	Users may access the services using a number of different terminal mechanisms and services, such as PCs, thin clients, plain old telephones, and/or wireless telephones.
Authorisation	Access may be authorised by multiple authorities, and authorisation may be delegated.
Revocation	Authorisation may be revoked at any time, and immediate enforcement of revocation may be required by policy.
Performance requirements	Response time requirements will vary according to application.
Event collation	Events for both security and management purposes need to be gathered, collated and managed.
Audit collation	Audit events from applications, middleware, and Web servers need to be collated and managed.

Figure 1.2 shows the functionality and mapping into the iTrust functional clusters. In the user-facing space, aspects such as single sign-on and VPN capability can be considered. These services interact with users through a client device into the iTrust functionality. In the authentication and authorisation areas, a number of technologies span the domains. Kerberos [3], for example, is involved in both the identification phase and the subsequent access-granting phase.

Looking at some of the currently available technologies, they can be placed within these functional clusters. The products and vendors shown in Fig 1.3 are illustrative only.

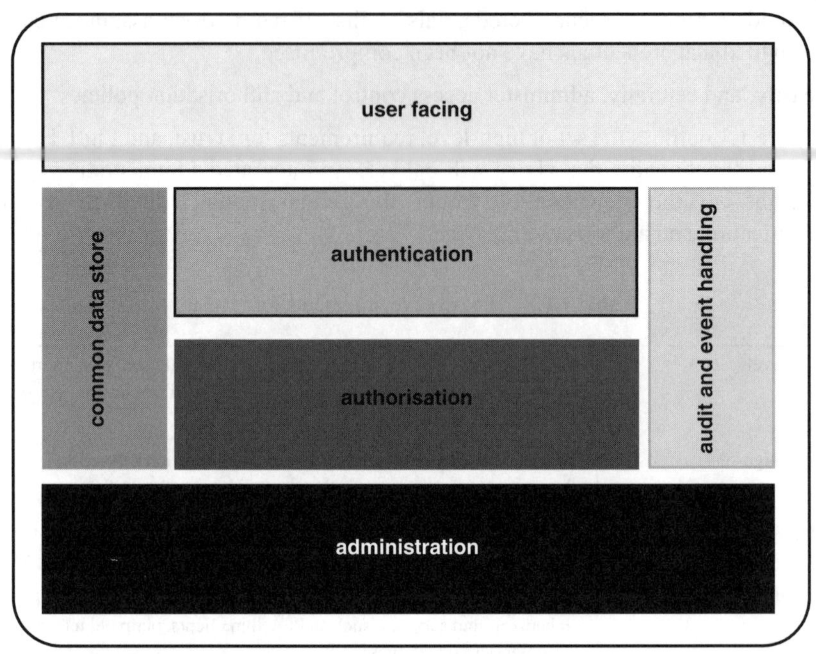

Fig 1.1 iTrust infrastructure functions.

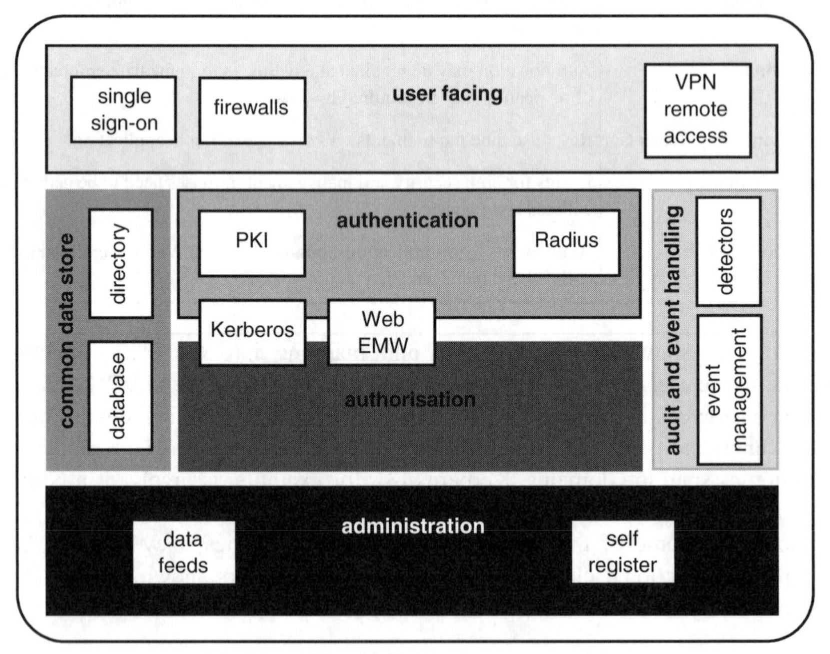

Fig 1.2 iTrust infrastructure components.

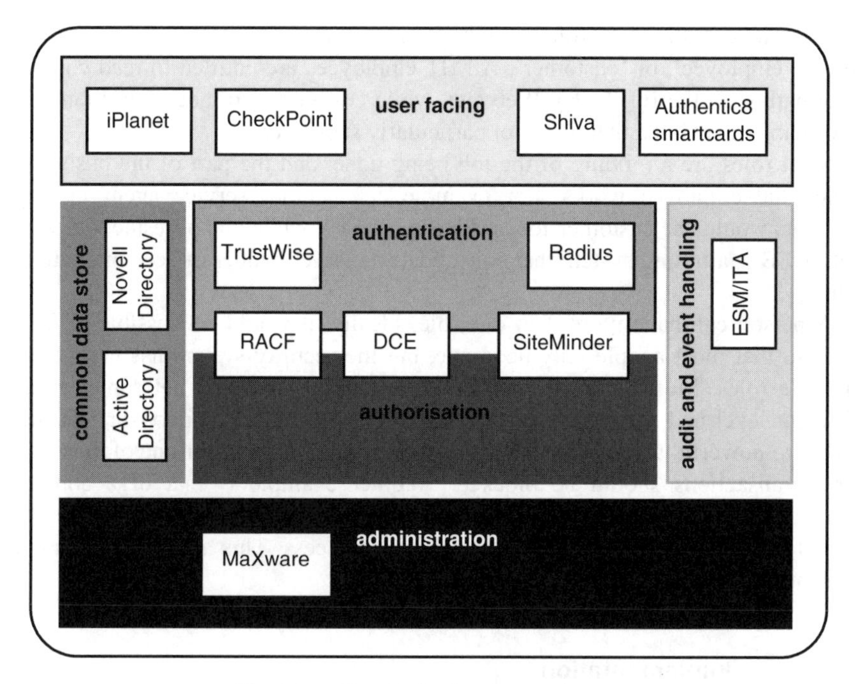

Fig 1.3 iTrust infrastructure packages.

1.5 Authentication and Authorisation

Essentially, access control security is split into two questions.

- Who is this person or device?
- Can this person or device access this resource in this way?

The first is authentication, and the second is authorisation. The data and timeliness of these decisions enables a number of other features which are also useful, such as audit trails, profiling, etc. This section describes the way in which BT has endeavoured to design the authentication and authorisation component.

1.5.1 People and Roles

Ideally, there should be a datastore for every individual, listing exactly what they are entitled to. In practice, this is unmanageable, and so the concept of a 'role' has been used. Role-based access control is generally believed to have first been outlined by Ferraiolo and Kuhn [4] as a counterpoint to the military multilevel security system described in the Bell-LaPadula model [5].

All people in the same role have the same access rights. A simple example of a role is 'employee', or 'customer'. All BT employees are entitled to read company information on the BT Today Web site, while customers are not, even though the information contained therein is not particularly sensitive.

Most roles are a mixture of the job being done, and the part of the business in which the employee works. For example, a customer service agent in major accounts would use customer-focused billing (CFB), while one in residential would not. (CFB is a billing system that consolidates a large number of accounts into one bill.)

A person can have more than one role. Ideally, it would be possible to define roles so that most people only need one, but the security system has to allow for multiple roles. Most roles are happy to live alongside others, but some are at a different level and so need to be exclusive. For example, a supervisor may have access to powerful transactions. When that person is filling in for one of their team, these transactions should be blocked. Another example is that of a computer administrator. Most of the time these people can do their work using the tools provided, but occasionally they need privileged access. This access requires extra authentication, and will be for a limited time.

1.5.2 Implementation

Even when the number of users has been reduced by collecting them into roles, there are still a lot of actions that each role can perform. The role is therefore divided into a number of parts, such that each part can be enforced independently. This also supports the security principle that no single person should manage the whole chain.

At present, a role is divided into three parts:

- the applications a person can access;
- the actions a person can do within each application;
- the data that can be provided to that person.

This division is done to enable the enforcement of the roles in multiple places.

1.5.3 Authorisation Components

Authorisation is divided into two parts — the enforcing function and the deciding function. For example, if a user tries to read a file, the file system is the enforcing function. It is the entity that actually provides the physical access. In the case of MVS, the deciding function is RACF, and the file system will ask RACF if the user can have access. In the case of simple Unix, the file system does the deciding itself by looking at the user's membership of groups, and the file's access rights from each group.

It is the intention to separate these components wherever possible and practicable. This separation will enable the number of decision components to be minimised and the economy-of-scale benefits gained. To do this, there needs to be an API between the enforcing and deciding components. At present, these APIs are proprietary (for example, MVS SAF — Security Authorisation Facility), but there are signs that standardisation will happen in the next few years. This is starting to happen in Java Authentication and Authorisation Service (JAAS) and Security Services Markup Language (S2ML), for example. Decision components need data to help them make their decisions. They need facts about the entity requesting access, and policies on when to grant access. These could be very simple — for example, a list of groups of which the user is a member, and a list of access rights for each group. Note that the enforcing component is a key part of the decision data, e.g. access to a database may be allowed through a specific transaction, but not directly via typed in SQL.

Where it is not practicable to separate the decision component from the enforcing component, better control can be attained by data sharing. This could be done by sharing a datastore between components, or possibly by populating an integrated component's datastore from a master datastore.

To simplify the administration, it makes sense to group resources into sets of related items, and then to give access rights to the set, not the individual items. For example, CSS groups its transactions into 'profiles', and then gives users access to profiles.

1.5.4 Handling Scale

Many solutions claim to scale. This is essential in a BT environment where one system alone (CSS) handles hundreds of millions of transactions per day. There have been several examples above where the problem of scale is being managed by grouping related things together (see Fig 1.4).

To create a business relationship between a person and a resource, several intermediate steps are created. Working backwards from the resource:

- the application designers (and business customer) collect related actions into action sets;
- the business process designers then collect action sets from several applications (and there may be more than one from each application) into a process (shown as profile 2 in Fig 1.4);
- the job designers (usually line managers) then organise their people to do a set of business processes (shown as roles in Fig 1.4);

- each person will have a number of roles, possibly from more than one manager — their collection of roles will make up their 'profile 1'.

It is therefore possible to construct security policies for the business out of the security policies for people, applications and business processes. This will greatly aid the understanding and verification of these policies.

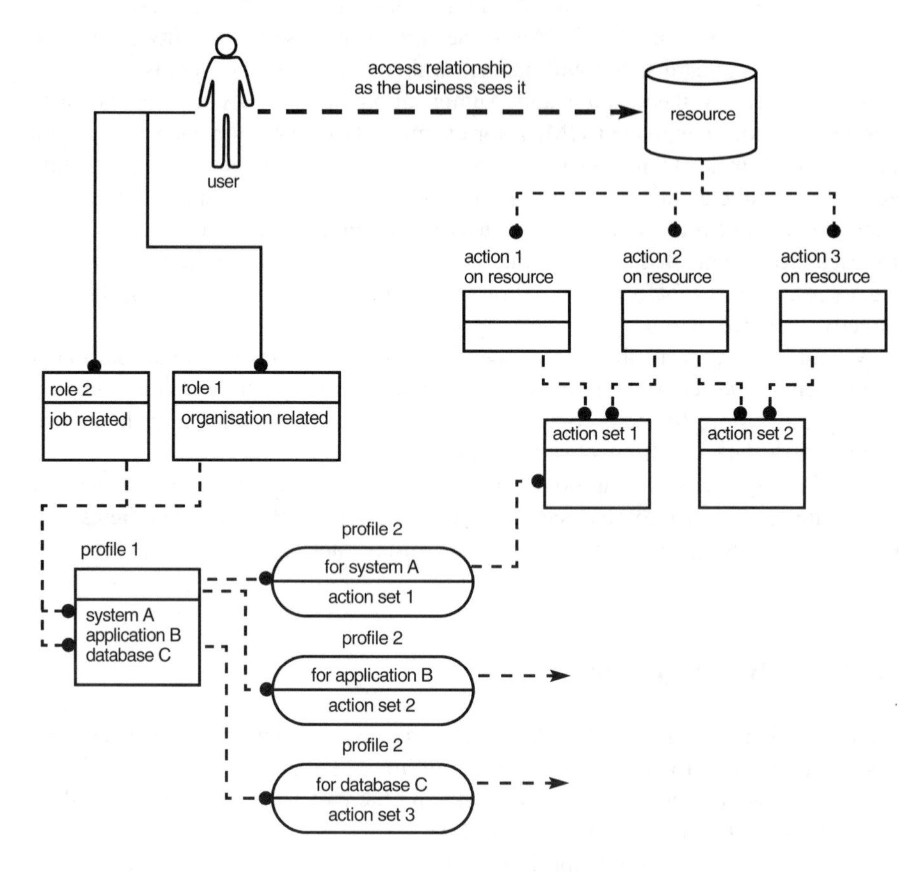

Fig 1.4 Access relationships.

1.5.5 Enforcement

Another reason for this separation is one of enforcement. Different software systems can be used to manage each of the above entities. A single sign-on product can enforce profile 1. Eventually, directory-enabled networks, for example using Microsoft Active Directory or virtual private networks, may enforce it. Actions and action sets are enforced within each application or resource manager. Profile 2 is ideal for storing in a directory or other common datastore. Note that this is only enforced by its use in defining who can access particular action sets.

1.5.6 Customers

Where customers access BT's systems, there is an additional requirement to ensure that they only act on and view the data belonging to their account. This is also true for BT employees in that they should only view and act on customer's data when customers make requests. This policy has been successfully enforced by manual procedures and audit checks. The sanction of internal discipline is adequate. Once the public has access to the systems, this method of enforcement will no longer work. The security systems therefore have to provide services to the applications that help to check that the end user is the owner of the data being accessed. This seems to be a simple association between an identity and the customer number of the data, but it does raise a number of queries about how the relationship is recorded.

However, it is clear that this issue should be solved once by the infrastructure, instead of being solved repeatedly by each application. Given an identity and a customer number, the security service should return the level of certainty that they represent the same (legal) person.

Customers accessing BT systems will do so via an external-facing portal. Users will log on and authenticate using one of the available methods appropriate for the services being accessed. Customer access credentials will not be mapped directly to back-end systems for subsequent validation. Instead, an agent task, acting on behalf of the customer, will undertake the transaction against the back-end system. This greatly simplifies the administrative task, since BT does not have to deal with being able to validate many millions of users from the back-end systems. However, to ensure end-to-end traceability, a mapping must be possible between the customer and the back-end system. Initially, this will be accomplished by audit records on the agent system, but ultimately by implementation of a primary correlation key (PCK) — an internal development to allow tracking of the originator of a transaction across multiple systems and audit trails.

1.6 Implementing the Architecture

As part of the architecture, authorisation and authentication are important features. BT has procured a product which seeks to externalise security from applications while also supporting single sign-on for Web-based applications and BT's enterprise middleware (EMW). This externalisation is a key feature that allows developers to make use of the infrastructure in supporting their system.

The delivery is made up of three pieces, all of which are explained in more detail below:

- application and Web server protection provided by SiteMinder;
- common directory;
- administration functions.

1.6.1 Application and Web Server Protection

SiteMinder (provided by Netegrity Inc) provides access control and authorisation capability for Web-based systems. SiteMinder is a directory-enabled, standards-based system that can work with nearly all BT's Web and application servers, operating systems, and application development platforms. SiteMinder:

- simplifies the development of applications — by removing the security implementation from the application and Web site developer, it allows developers to focus on their task, with security becoming a configuration task;
- has flexible configuration — the control of access to resources can be delegated to any administrator;
- provides scalability for an enterprise, as it is able to support millions of users and thousands of applications — deployed with an appropriate access control model, it can cope with the scale of BT.

1.6.2 Directory Integration

Standardisation is an important aspect of products that BT procures. It provides the flexibility of integration and avoids supplier 'lock-in'. LDAP (lightweight directory access protocol) [6], which SiteMinder supports, has been identified as being a crucial standard in the deployment of access control systems. SiteMinder provides further flexibility in that authentication may be done against one directory and authorisation against another. BT has integrated SiteMinder with an LDAP directory holding all 130 000 employees. As the company federates, SiteMinder will be integrated with the various directories in the new businesses.

SiteMinder is very flexible in the use of directories and the data within them. It is possible to implement dynamic look-ups from business data during authorisation which permit:

- customisation of the user experience;
- further relevant information being automatically presented to a Web application on behalf of the user;
- further decisions on access entitlement to be made.

SiteMinder has been integrated with the BT directory service to provide user authentication, and to enforce access control policies based on a user's identity attributes and group membership. SiteMinder uses the directory in several ways:

- users are authenticated against their identity in the directory;
- policies are held within the directory — when a user attempts to access a protected resource, all the policies associated with protection of that resource are checked to determine if they apply to the user;
- personalisation of a user's experience can be accomplished by including user attributes from the directory in SiteMinder responses.

1.6.3 Administration

SiteMinder provides a centralised approach for undertaking security management. The centralised approach offers several advantages.

The time and cost of developing and maintaining multiple systems is reduced, making it similar to managing one security system. This is achieved by using a consistent security policy across all the Web applications that are to be protected. This consistency and centralised approach has the following benefits:

- security functionality does not have to be written in a bespoke manner for each individual Web application;
- the time and cost to deployment is reduced, since only one central security system is being manipulated.

The user interface for configuring policies is a Web-based application. Administrators configure policies and the responses that supply data to Web applications. Administrators can update policies as the needs of the business and the user community change. The SiteMinder rules can be adjusted to establish a set of permitted actions on a resource, and for that set of actions to be mapped to user groups.

1.6.4 Components

The SiteMinder architecture (see Fig 1.5) is based around two components:

- policy server;
- Web agents.

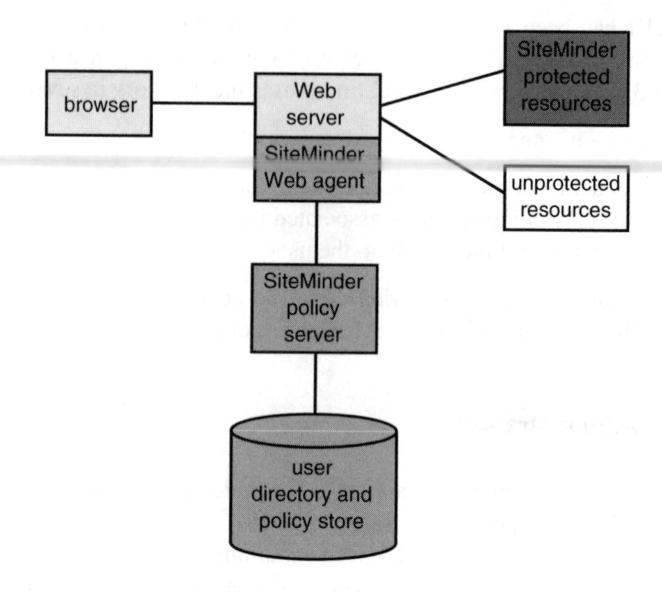

Fig 1.5 The SiteMinder architecture.

There is no specific client-side component required. The system uses a standard browser without any modification.

The policy server is the hub of the system. It obtains information on users and resources protected from the user directory and policy store. The policies are delivered securely to the Web agents as required, and these policies are cached by the Web agent for performance reasons. Web agents are available for a number of the leading Web servers. The agents integrate with the Web server, application server, or custom application, to enforce access to resources based on the policies. Web agents can control access to pages, CGI forms and script parameters.

Audit logs are generated by the policy server relating to user activity. These logs can be printed as predefined reports for analysis of breaches of security policy.

For sites that cannot be protected with SiteMinder, for example because the site is outside the BT domain, or there is no Web agent for the particular Web server technology, the solution is to use a reverse proxy. In this configuration, clients would access the reverse proxy where that access control decision would be made. The reverse proxy would then pass on the authorised request to the site.

It is possible to secure resources that are held on application servers conforming to the Java 2 Enterprise Edition (J2EE) standard. These resources may be Servlets, Enterprise JavaBean (EJB) components, or Java Server Pages (JSP). This is important since it allows BT to protect substantial amounts of its planned middleware resources (see Fig 1.6).

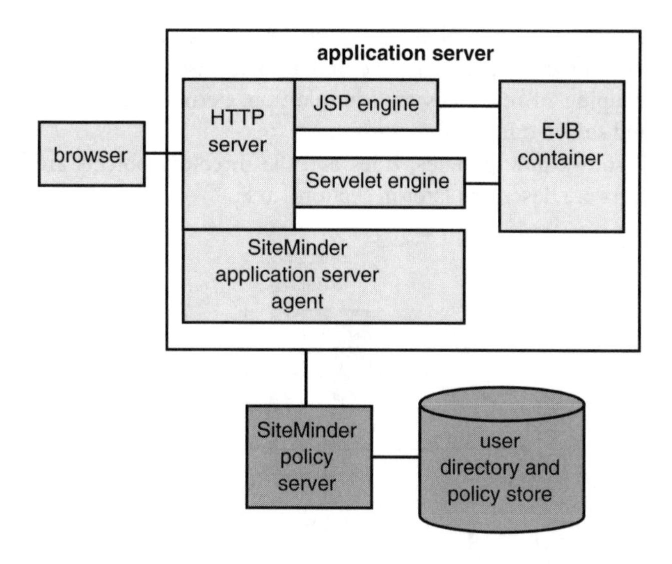

Fig 1.6 Reverse proxy process.

1.6.5 Policies and Resources

Policies protect resources and are able to either allow or deny access. Policies specify the resources to be protected, the groups of users and the conditions under which access is to be granted. It is possible to indicate within the policy how a user is to be treated in the event of an authentication failure. For example, it would be possible to redirect the user to a page where the user could submit an access request, or alternatively to a page stating 'Access denied'.

Other restrictions may be associated with policies, such as the requirement to be from a particular IP address, or accessing the site within particular time periods.

There are many types of resource to which a user may attempt to gain access. Table 1.2 shows some examples.

Table 1.2 Types of resource.

Web page	/www.company.com/index.html
CGI script	http://finder.company.com/webferret/ search?query=skills
Directory	/mydirectory
Servlet or EJB	Company.servlet.app1
ASP page	http://finder.company.com/webferret/ search.asp?query=skills
JSP page	/mydirectory/app2.jsp

SiteMinder uses resource filters to specify resources to protect. They provide groupings of resources and reflect the company's infrastructure organisation. They permit the grouping of resources and, in addition, permit specific resources to be highlighted and selected for different users.

Figure 1.7 shows how policies, rules and the directory objects are linked. The directory objects are described later in section 1.6.8.

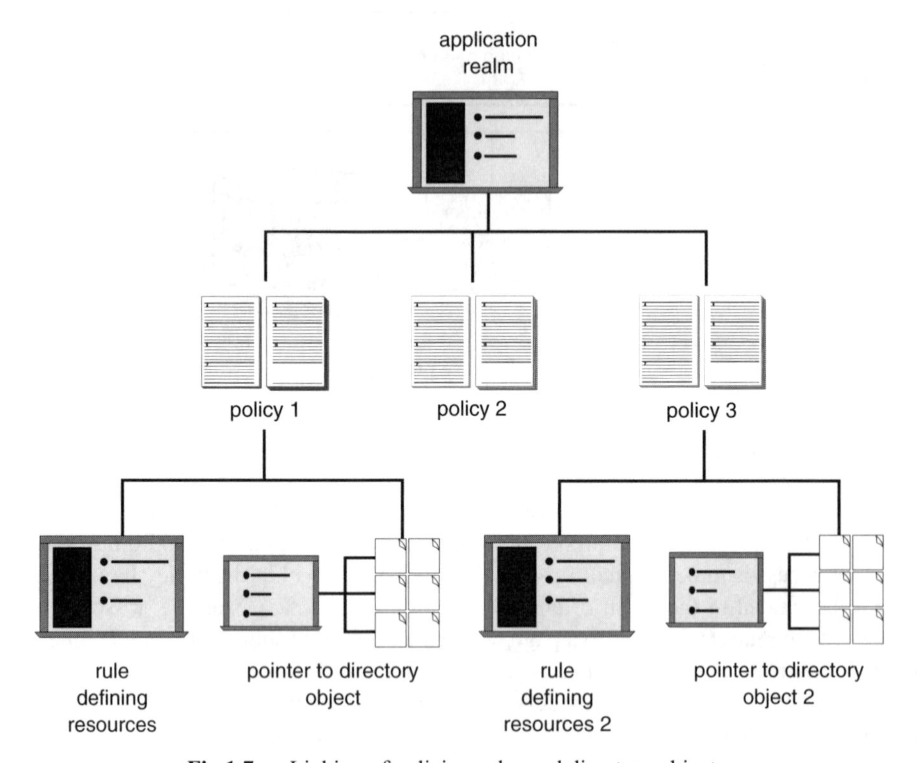

Fig 1.7 Linking of policies, rules and directory objects.

1.6.6 Single Sign-on and Authentication

Single sign-on describes the concept of identifying an individual once, and then, having identified the individual, permitting access to controlled resources without further proof being required from the user.

With complex configurations SiteMinder needs to be configured to manage specific domains, e.g. intra.bt.com and nat.bt.com. SiteMinder permits the user to move between domains without any further interaction on behalf of the user.

In a complex environment, there will typically be resources that require more protection than others. For example, an organisation may require simple 'UserID and Password' for access to some administration systems, but may require stronger controls or proof, such as smartcard or biometrics, for financial transactions. Protection levels may be assigned to resources which can indicate the required degree of authentication that is required.

1.6.7 Scalability and Performance

When a great deal of dependency is placed upon infrastructure components it is clearly important that resilience is built in. BT will have a number of policy servers located at strategic points around the country.

It is possible to configure how traffic is managed across replicated systems. Load balancing and fallover is supported between policy servers, Web agents and directories.

1.6.8 Common Directory

The directory is shared with other security functions and hence it provides a common point of definition of people's access rights.

iTrust will provide a single logical directory, replicated for resilience, to hold details about all BT employees and contractors. This directory can also be used to hold information on suppliers and customers.

The directory is structured as a tree (see Fig 1.8). At the top is BT, and, below that, namespaces for employees, roles and applications. It is very straightforward to add branches to hold other objects, such as customers.

Users are assigned membership of groups. These groups are currently designed to represent various roles within an application, but, over time, they will be generalised to represent job-related roles. These roles will then be associated with several applications.

Each application is associated with a container object, and that object has a set of containers within it to represent the various policies defined in SiteMinder. This direct association of policies and directory containers will make it clear what the security requirements are for any resource. The groups of users will be put into these containers and hence users will be linked to a policy and thus given access to a resource.

This model is general, and can be defined in terms of LDAP queries for applications that can work directly with a directory. It also makes it easy to export the details of users with access to an application as a flat file, for import into other systems.

The user object can be used to hold other security-related data, e.g. for storing information to support legacy single sign-on.

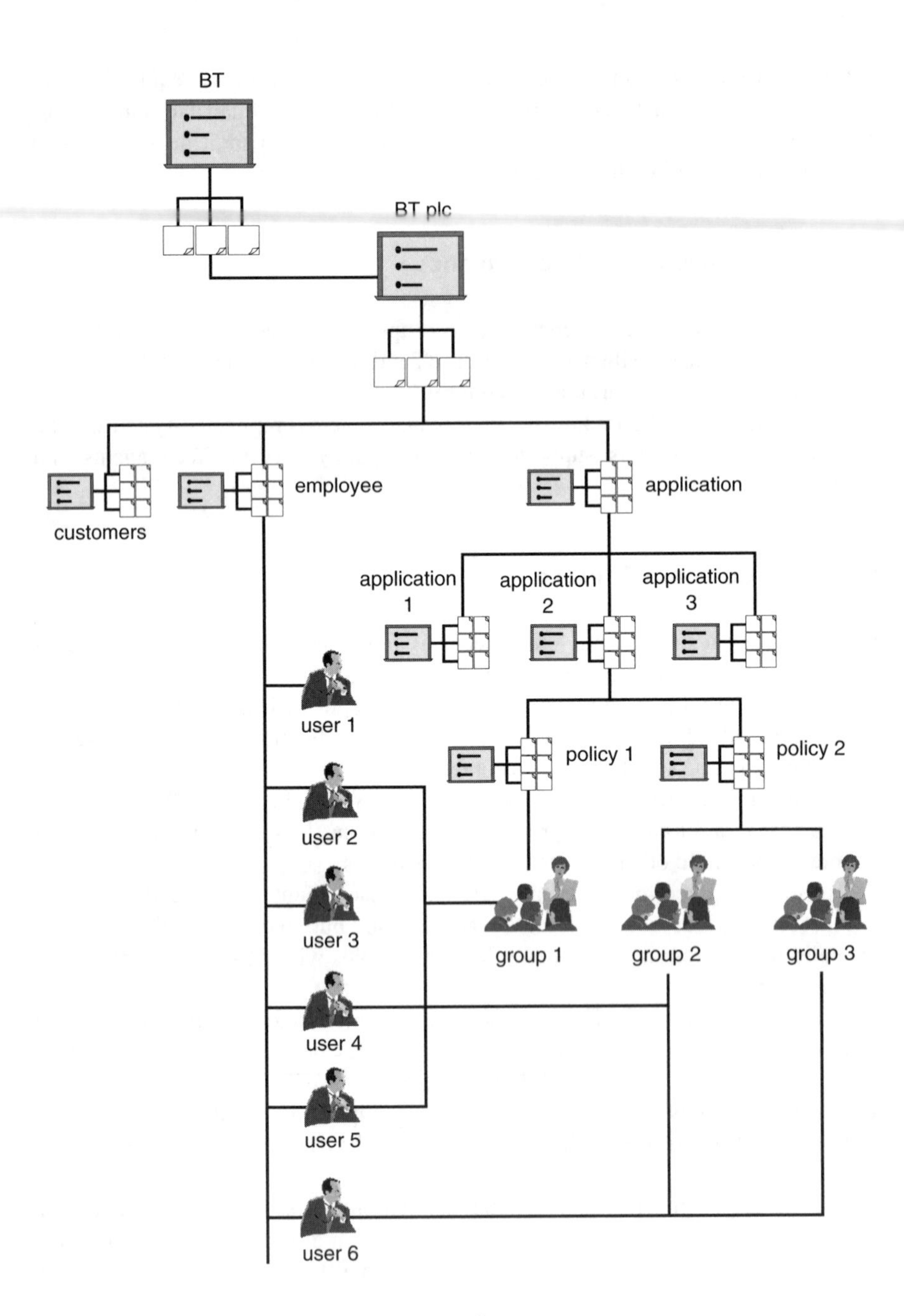

Fig 1.8 Common directory structure.

1.6.9 Administration Functions

The administration of application and Web server security is divided neatly into two parts by the above design for policies and directories. The first part defines resources and links them, through a SiteMinder policy with a container object in the directory. The second part defines users and groups, and links them to the same container object.

The definition of resources and policies is done using the SiteMinder user interface. This has been described in section 1.6.3 above.

The definition, addition, removal and modification of employee users is handled by the authoritative systems within BT, and feeds are taken from these on a daily basis. The main source systems are HR (PeopleSoft), Group Finance (eOrganisation) and TeamConnect (ODS/Exchange). This process will be automated, making a significant reduction in administration costs. It also means that applications can rely on the data in the directory. If users exist (and are not disabled), then they will be current employees.

A similar link to authoritative systems will be needed for business customers. Initially, while the numbers are small, this can be manual. An automated system will be needed once the numbers grow — and a sensible option from day 1, if the system is used by consumers.

1.6.10 PKI and Digital Certificates

A critical aspect of eCommerce is to establish trust between trading entities. This trust covers confidence that what was sent was actually what was received, that the information was actually received, and that the sender cannot later deny that the information was sent. A technique for providing this is the use of digital certificates, supported by a public key infrastructure. The iTrust architecture and design supports the use of certificates and BT is currently using them in a number of contexts, including secure e-mail and remote access via public switched networks into the corporate network. For BT people, the target is to hold these certificates on smartcards which will allow both remote access from the Internet and also access to resources from the company intranet. This will become a reasonable proposition when card readers are ubiquitous and do not add significantly to the cost of the basic machines.

Of course, there are additional costs with running a system with smartcards and certificates, including running revocation, issuance and registration authority functions, and also certificate renewal. These require careful consideration in any business case although the cost savings associated with reduced password resets must also be factored in.

1.7 Summary

BT is currently splitting itself into separate lines of business (LoBs) and there will be legal and regulatory controls over who has access to what information — as well as commercial implications. Some customers of one LoB may also be the direct competitors of another LoB. In addition, there are substantial technology integration challenges to be faced, both in building up new systems, and infrastructure, yet at the same time decoupling systems, where it is appropriate, to fit into the LoB model.

The iTrust infrastructure is shaping up well to these challenges, although as with many things in this complex technological world, implementation must be seen as a journey, not necessarily something that has a clear and distinct end. The vision is known, as is what has to be done to reach it. Getting there, however, will be influenced by changes in the business, new technologies, ability or desire to integrate legacy systems and, of course, the risk profile.

References

1 HMSO — http://www.legislation.hmso.gov.uk/acts/acts/1998/19980029.htm

2 Institute of Chartered Accountants in England and Wales — http://www.icaew.co.uk/news/pressoffice/

3 Kerberos (IETF) — http://www.ietf.org/rfc/

4 Ferraiolo, D. and Kuhn, R.: '*Role-based access control*', in Proc of 15th National Computer Security Conference, NIST (1992).

5 Bell, D. E. and LaPadula, L.: '*Secure computer systems*', Vols I—III, Mitre Corporation (1973—1974).

6 LDAP (IETF) — http://www.ietf.org/rfc/

2

XML AND SECURITY

A Selkirk

2.1 Introduction

Over the past twenty years, there have been security standards created on a wide range of subjects, from constructing digital signatures to formalising access control mechanisms. Standards groups such as ISO [1] and the IETF [2] have constructed security protocols for solving common security problems involving communicating networked machines. Many of these standards are being reconstructed using XML as a syntax. Understanding why this process is taking place and why it is important is not easy. It requires an understanding of XML, a technology originally intended to solve certain Web browser issues. The positioning of XML as a key business technology necessitates security standards to safeguard transactions. Constructing these standards in XML allows better interoperability with application data. However, that is only part of the story — XML can also be used to construct new security mechanisms with better functionality. This chapter explores XML, XML Signature, and the standards that are being constructed to make use of them.

The first two sections provide background history and a quick explanation of XML, which can be skipped over by those who already understand XML. The XML Signature [3] standard is presented in depth, explaining the novel features that it employs. The XML family of technologies includes many that are used by XML Signature (and the XML security protocols), such as XML Schema, XML Namespaces and XSLT. How XML can utilise encryption mechanisms is discussed.

2.2 A Brief History of XML

Should flashing text be a common HTML tag? This was a subject for debate in the early years of the World Wide Web. From its beginning in 1990, the markup language used to display pages on the World Wide Web was HTML. One reason for the success of HTML was its simplicity. It was a text format, and could be created using any simple text editor. During the 1990s, Microsoft and Netscape battled to create the best Web browser, and each side introduced new HTML features to provide a technological advantage over their rival. So tables, frames and other not-so-well-received ideas began to complicate HTML in advance of official standards. Potentially a Web page could make use of features as diverse as mathematical symbols, vector graphics or sound. An official HTML standard was necessary for Web browsers to interoperate, but incorporating all possible features into HTML would have made it huge and hard to comprehend [4]. It would also be difficult for Web browsers to comply fully with the standard. On the other hand, if HTML was kept small, then little improvement on a simple 'text and images' Web would be possible.

Eventually the rival companies agreed to a solution. Today, under the W3C standards group, HTML has been standardised around a common set of text and graphical features. New markup languages which contain specialised functionality are now created under a common syntax. This syntax is called XML, the Extensible Markup Language [5] — this is of course a very narrow viewpoint on the history of XML, but it illustrates a major reason for its existence. The creators of XML aimed to make it the common syntax for a number of Web display languages as well as for the data that Web pages would display. Unexpectedly it has become the foremost business data format, and is branching out into computer protocols and even security mechanisms. Both HTML and XML have their roots in SGML, the grandfather of current markup languages. Where SGML failed was in its complexity, which meant that it was difficult to implement in its entirety. XML was designed so that it would be easy to create XML documents, and easy to create programs to process these documents.

The XML standards have been created by the World-Wide Web Consortium (W3C). Tim Berners-Lee, the original creator of the World Wide Web, is the Director of the W3C and most of the leading technology companies (including BT) are members. There are many important standards created by the W3C, including XML itself (completed in 1998), and HTML 4.0 (1997). The Document Object Model (DOM level 1: 1998 [6], DOM level 2: 2000) describes XML and HTML in a tree model, allowing client-side processing of Web pages using JavaScript or Java applets. XML Stylesheet Language Transformation (XSLT), completed in 1999, provides a syntax for transforming an XML document into another XML document, or HTML, or even an alternative text format. XSLT is mainly used for transforming data held in XML format to an HTML display of the data.

There are five stages in a standard's progress at the W3C. A W3C note is a proposed standard submitted by a W3C member but not endorsed in any way by the W3C. It can be used by groups to pave the way for an official standard to supersede the note. For example, before the XML Schema standard was begun, a number of XML Schema languages were submitted (XML-Data [7], Document Definition Markup Language [8], Schema Object-oriented Language [9]) to provide models for early implementations and to propose how the official language should work. If there is sufficient interest in creating a standard in a particular area, a working group is set up and its objectives established. A Working Draft is a standard that is still undergoing major changes. A Candidate Recommendation is a standard that is considered complete in essentials, but is looking for actual implementations to ensure that it is workable. A standard reaching Proposed Recommendation status has been completed by the working group and awaits endorsement by the director of the W3C. Finally a stable standard endorsed by the Director is called a Recommendation.

Other standards groups interested in XML are the Internet Engineering Task Force (IETF), and the Organisation for the Advancement of Structured Information Standards (OASIS). The latter is of especial interest to the security community, because it has a security services technical committee, creating the important XML security standard, the security assertion markup language (SAML).

2.3 An XML Primer

Flashing text did not become a standard HTML markup tag. Neither is it an official XML tag either. In fact there are no official XML tags — XML is a syntax for creating markup languages, not a markup language itself. XML is a text-based standard, and can usually be read by a human. A short example is shown in Fig 2.1, easily understandable without even the slightest knowledge of the situation.

```
<?xml version="1.0" encoding="UTF-8"?>
<customer>
  <name>Fred Smith</name>
  <address country="UK">
    <street>93 Example Avenue</street>
    <city>Ipswich</city>
    <region>Suffolk</region>
    <postcode>IP50 9BT</postcode>
  </address>
  <status purchases="2" last-purchase="12-05-99"/>
</customer>
```

Fig 2.1 Short example of XML.

The angle brackets (< and >) denote tags. There are two types of tag — a start tag and an end tag. Inside a start tag there is a (usually self-describing) word, called the **element** and optionally a number of **attributes**. An attribute is a pair of words consisting of a name and a value and written in the form: x = "y", where x is the name and y is the value. Therefore, in the example, the **elements** are 'customer', 'name', 'address', 'street', 'city', 'region', 'postcode' and 'status' while the **attributes** are 'country', 'purchases' and 'last-purchase'. An end tag contains a forward-slash and an element name that matches the element of the start tag. Each opening tag has a complementing closing tag with the same word between the angle brackets except for the forward-slash at the beginning. Between start and end tags there can be text known as **character data**, or other tags, or both. In the example in Fig 2.1, the character data is 'Fred Smith', '93 Example Avenue', 'Ipswich', 'Suffolk' and 'IP50 9BT'. A tag containing neither child elements nor character data can be written in an abbreviated form <tag/>. This is the same as <tag></tag>. In Fig 2.1, the element 'status' uses the abbreviated form.

An XML document always begins with the XML declaration `<?xml version="1.0" ?>` which states the version of XML being used. This is to allow future versions of XML to be treated differently. The declaration can optionally announce the character encoding system that is being used, based on the different varieties of Unicode that can be used.

The XML will be read by an XML parser. A parser is an application that can create from the XML the data structures that the XML represents. XML data is often modelled as a tree structure (see Fig 2.2).

```
customer
  |
  +--name: Fred Smith
  |
  +--address (country=UK)
  |    |
  |    +--street: 93 Example Avenue
  |    |
  |    +--city: Ipswich
  |    |
  |    +--region: Suffolk
  |    |
  |    +--postcode: IP50 9BT
  |
  +-- status (purchases=2, last-purchase=12-05-99)
```

Fig 2.2 XML modelled as a tree structure.

In this case, 'customer' is the root XML element, with child elements 'name' and 'address'. 'address' in turn has child elements 'street', 'city', 'region' and 'postcode'. The Document Object Model (DOM) describes XML documents as such a tree structure of elements, attributes and character data.

Tags must be nested or disjoint — if a start tag is nested within the content of another tag, then the end tag must also be nested, before the end-tag of the outer tag (see Fig 2.3). This contrasts to HTML, which is more laid back on how tags relate to each other, and does not require end tags for certain types (the paragraph tag, for example).

```
<tag1>Hello <tag2>There!</tag2></tag1>
  [A well-formed XML fragment]

<tag1>Hello <tag2>There!</tag1></tag2>
  [Illegal due to overlapping tags]
```

Fig 2.3 Examples of nested tags.

XML is a reasonably simple standard, and the essential parts can be learned quickly. It is more structured than HTML — for example <P> is a different tag to <p> and unlike HTML each opening tag must have a closing tag. This makes interpretation of valid XML easier, since there is no leeway for ambiguity. It is not tolerant of errors — either XML is well-formed or it is not valid. This is a quick introduction to the more common features of XML, but it is not complete and more comprehensive information can be found elsewhere [10].

2.4 The Benefits of XML

There is a well-known joke that XML is superior because the angle brackets (< and >) help streamline the data as it passes through the wire. In reality there is no inherent advantage in the syntax of XML over similar markup formats. Its appeal lies in its universal use, and that tricky low-level technical problems, such as the use of different alphabets, have been solved. Due to XML being text based, an XML document is easy for humans to understand. An XML format is also easy to extend and integrate with other formats. Extensibility arises from the redundancy within text — it is easy to create new element and attribute names that do not interfere with the existing name set.

The numerous tools for parsing and processing XML and the standard models for representing the parsed data mean that a programmer has a head start in using XML data over other text APIs. There is also a great advantage in having many different standards with a common format. Standards can build on or fit into previous XML standards. Because it was explicitly created for Internet use, it can use existing technologies such as HTTP (the World Wide Web protocol) and easily transform into HTML for display on current Web providers. Finally, as there is an easy upgrade path from HTML to XHTML and XML technologies, future Web applications are not reliant on a technology that may not succeed.

The advantage that XML brings to security standards is integration with other XML data vocabularies. As an illustration, an authentication protocol in XML can contain information from another XML vocabulary. There is also an advantage of different security protocols using an XML rather than separate binary formats, since they can better understand and work with each other.

2.5 XML Technologies

The XML standard was completed in 1998. Since then, there has been a long list of markup languages using its syntax. There has been the obvious recasting of HTML to XML syntax, known as the XHTML standard. Mathematical symbols can be represented in XML using MathML. An example of the spread of XML to more specialised areas is Meatxml [11], a proposed standard for meat and poultry trading.

There are efforts to enhance the functionality of the original XML standard so that it can be used for more complex purposes — for example, the introduction of data types. Specific functionalities such as hyperlinks between XML documents and the transformation of XML formats have been considered and are in use. XML is also used by a large number of business data standards. The new XML security standards will take advantage of these technologies to achieve better functionality, security granularity, or ease of use. Therefore the potential power and purpose of security mechanisms using XML can only be understood by examining the family of XML standards.

2.5.1 Extending XML Functionality

2.5.1.1 XML Namespace

In creating large numbers of XML formats there need to be ways to avoid name clashes while allowing formats to work together. The XML Namespace standard [12], created in 1998, allows XML documents utilising vocabularies from more than one standard to work coherently, understanding which element comes from which vocabulary. This is important especially for standards that need to make use of previous standards — such as the security services standards that will use the XML Signature standard. For example XHTML has a 'table' element for tabular display of text. A hypothetical furniture vocabulary may use 'table' for a particular furniture item.

Using XML Namespace, elements and (optionally) attributes are given prefixes indicating the vocabulary (or namespace) that they come from. A namespace is declared in an XML document using an attribute in the form:

```
xmlns:*="urn:namespace"
```

where `*` represents the prefix and `urn:namespace` represents the namespace unique resource identifier. Using the `xmlns` attribute on its own defines the default, prefixless namespace. Examples of namespace declarations are:

```
xmlns:dsig="http://www.w3.org/TR/Signature"
```

and

```
xmlns="http://www.w3.org/TR/2001/XMLSchema".
```

The first declaration states that all elements and attributes that are prefixed by 'dsig' belong to the digital signature namespace, while the second states that prefixless elements and attributes belong to the schema namespace.

Figure 2.4 shows a complete example of namespaces in action, taken from the S2ML standard.

An element or attribute with a prefix is of type QName. QName stands for qualified name and represents an XML element or attribute which contains a prefix from a recognised namespace.

```
<?xml version="1.0"?>
<Entitlement xmlns="http://ns.s2ml.org/S2ML"
    xmlns:dsig="http://www.w3.org/2000/09/
    xmldsig#">
  <ID>urn:financeDepartment:129de12</ID>
  <Issuer>http://www.example.com/finance/
  AzEngine</Issuer>
  <Date>2000-10-16T12:34:120-05:00</Date>
  <AzData>
    <SC:PaymentRecord xmlns:SC="http://ns.finance
    -vocab.org/finance">
    <SC:TotalDue>19280.76</SC:TotalDue>
    <SC:Over60Days>1200.00</SC:Over60Days>
    <SC:Over90Days>10000.00</SC:Over90Days>
    </SC:PaymentRecord>
  </AzData>
  <dsig:Signature> ... </dsig:Signature>
</Entitlement>
```

Fig 2.4 S2ML standard namespaces in action.

The example in Fig 2.4 shows three namespaces working together. The main namespace (`http://ns.s2ml.org/S2ML`) is from the S2ML standard, which defines authentication and authorisation tokens. Here there is an entitlement issued by a certain authorisation engine for a certain time period. The second namespace (`http://ns.finance-vocab.org/finance`) declared in the PaymentRecord element is a hypothesised finance vocabulary, which describes the payment record of the entity. The third namespace is the XML digital signature namespace. The digital signature (details of which have been omitted from the example) provides the security binding necessary for the token to be accepted as genuine. The S2ML standard does not wish to define a definitive vocabulary for authorisation information, so it allows vocabularies from other namespaces. Similarly, it does not want to define digital signatures and therefore the signature part is taken from the W3C XML Signature standard.

2.5.1.2 DTD and Schema

The XML standard was created with an associated validation language, the document type definition (DTD). This allowed the basic structure of an XML document to be defined. How elements and attributes fit together is an essential part of XML because it allows the creation of standard formats with a describable structure. Although widely used, the document type definition has been considered insufficient for several reasons. Firstly, it uses a non-XML format, so that a separate DTD-parser is needed to parse DTDs. Secondly it does not specify data types for attributes and character data. The XML standard does contain a few basic types, but this is to help describe its format, not for use in XML attributes and character data. A DTD cannot require that an attribute value is numeric, or a positive integer, or a list of country names. Thirdly a DTD does not easily integrate with XML namespaces and multiple XML vocabularies. Finally there is no way to reuse the definition of an element when specifying a new element with extended or restricted functionality.

XML Schema [13] is the W3C standard that aims to rectify this situation. As of May 2001, it has become a full W3C recommendation, indicating a mainstream, stable standard. It is a large standard, split up into two parts (schema structure and schema datatypes, both around 100 pages long), with an explanatory Primer to ease the learning curve. Part of the schema for XML Signature, showing how the top-level "`Signature`" element works, is given in Fig 2.5.

```
<?xml version="1.0"?>
<schema xmlns="http://www.w3.org/TR/2001/XMLSchema"
  xmlns:ds="http://www.w3.org/2000/09/xmldsig#"
  targetNamespace="http://www.w3.org/2000/09/xmldsig#">
[...]
<element name="Signature" type="ds:SignatureType"/>

<complexType name="SignatureType">
<sequence>
  <element ref="ds:SignedInfo"/>
  <element ref="ds:SignatureValue"/>
  <element ref="ds:KeyInfo" minOccurs="0"/>
  <element ref="ds:Object" minOccurs="0"
  maxOccurs="unbounded"/>
</sequence>
<attribute name="Id" type="ID" use="optional"/>
</complexType>

[...]
</schema>
```

Fig 2.5 Part of the schema for XML Signature.

This shows that the "`Signature`" element has four possible child elements, sequentially ordered. The first two (`SignedInfo` and `SignatureValue`) are mandatory, the last two (`KeyInfo` and `Object`) are optional (`minOccurs="0"`), and `Object` can be repeated a number of times (`maxOccurs="unbounded"`). `Signature` also has an optional attribute, `Id. SignedInfo, SignatureValue` and all other elements will in turn be specified by the schema, so that the whole structure of the Signature format will be known. Figure 2.6 shows an XML fragment conforming to this schema fragment. (Note that `xmlns` is not considered to be a conventional attribute — it does not need defining by the schema.) The child elements arising from `SignedInfo`, `SignatureValue` and `Object` are not included.

```
<Signature xmlns="http://www.w3.org/2000/09/xmldsig#">
  <SignedInfo>[...]</SignedInfo>
  <SignatureValue>[...]</SignatureValue>
  <Object>[...]</Object>
  <Object>[...]</Object>
</Signature>
```

Fig 2.6 XML fragment conforming to the schema in Fig 2.5.

XML schema can perform a separate valuable security task — ensuring that data inputs and outputs are in a proper, expected form. Buffer overflow and similar attacks may be preventable by the careful creation of a schema which limits string lengths or enforces strict maximum values for integers. The schema validation process can then reject XML documents that do not obey these restrictions.

XML that conforms to a DTD is called validated XML, while XML that conforms to an XML schema is called schema-validated XML. Already the more up-to-date XML standards such as the simple object access protocol (SOAP), XML Signature and XML Key Management Specification (XKMS) use XML schemas to define their structure and data types.

2.5.1.3 Parsing XML

A text document in an XML format is not useful by itself. There needs to be some method by which it is turned into actions or information on a computer. The computer applications that perform this are called parsers. There are several standards for the representation of an XML document after parsing and its interface to an application. The document object model, already mentioned, creates a tree-model of element, attribute and character data objects. The simple API for XML (SAX [14]) sees elements, attributes and character data as events, allowing a programmer to handle each element or attribute as it is read. It is likely that there will be new APIs in the future that will incorporate XML schemas, allowing the type of an element or attribute to be taken into account. The W3C are currently working on a standard, the XML Infoset, that details the information that a parser needs to take from an XML document and provide in its API.

2.5.1.4 Display

Where human decisions have to be made, there needs to be the graphical representation of data. Although HTML still rules in the Web browser world, there are a number of XML languages that should win out in the longer run. The simplest is XHTML, a reformulation of HTML in an XML syntax. The number of changes is not high, since HTML is close to XML in syntax. It is possible to write XHTML Web pages that can be accepted as HTML by most Web browsers.

In the future, Web browsers should display more than the formatted text, tables and images that is HTML. Two candidates for incorporation in Web browsers are MathML and SVG. MathML is the XML format for displaying mathematical symbols and equations. This may in the future supersede LATEX as the format for mathematical papers. SVG (scalable vector graphics) is a format for describing simple geometric shapes. This could enable Powerpoint-like diagrams on Web pages rather than images.

Where XML is being used in a big way for display is the mobile telephone market. Technologies such as WAP and i-mode make use of simple markup languages such as WML and CHTML to display text. There is also an XHTML-Basic standard, which takes the smallest possible subset of XHTML for use in future simple display situations.

Display is important for digital signatures when human interaction is needed. Whether there is data that has been signed or data to sign, a faithful visual representation of the data will enable a human to decide whether to accept (or create) a signature. The information within a signature may also need displaying. If the display format and the signature format both use XML, the conversion process from data to visual information is easier. It also makes signing visual information straightforward.

2.5.2 Transforming, Hyperlinking and Traversing XML

A major reason for creating XML was to separate data that a Web page may contain from its presentation. The idea was to create a stylesheet language to transform the XML data into an XML display format. The original XML Stylesheet draft specification was split into two parts — XSL Transformations (XSLT) and XSL-Formatting Objects (XSL-FO or just XSL). XSLT became a W3C recommendation in November 1999 and has been hugely successful. It is a transformation language mainly used for converting XML to display formats such as HTML or WML, but it has an unforeseen secondary use in transforming one XML data format to another, otherwise incompatible, format.

A necessary requirement for the XSLT standard was a vocabulary for specifying parts of an XML document, to indicate what was being transformed. A separate standard, called XPath, was created to provide this. XPath sees an XML document as if it were a file directory, since both have tree-like structures. It is a vocabulary for navigating and referring to other nodes (elements, attributes, text and so on), starting at a specific node. XPointer is a draft standard extending XPath to create a more comprehensive model, including referring to parts of an XML document from another document. XLink is a proposed standard for extended hyperlink capabilities, so that XML documents can refer to or use other XML documents. XPath, XPointer and XSLT are used by XML Signature to refer to or transform the XML documents that are to be signed.

2.5.3 XML Protocols

There is an increasing number of XML formats for communicating between various parties. The best example of this is the SOAP format, which creates simple client/server communication. Jabber is an open source instant-messaging protocol written in XML, with the potential for extended uses such as conferencing and collaborative applications.

2.6 XML Digital Signatures

While the advantages of incorporating the results of esoteric mathematical algorithms using very large numbers into a text format are not immediately apparent, the W3C have been working on just such a task.

Beginning in the middle of 1999, the XML Signature Working Group has been creating a specification for defining digital signatures in an XML format. The most recent version of the draft standard, published on 31 October 2000, is at Candidate Recommendation status — requiring implementations for feedback.

There are two reasons for an XML signature standard when there are alternative mechanisms such as SSL for protecting data in transit.

Firstly, an XML signature, entwined within the XML data, is portable. It is a constituent part of the data rather than a sub-layer stripped off as the data is retrieved from the network.

The second reason is flexibility — XML signatures can refer to many documents or parts of a single document, as is required (Fig 2.7). Since the definition of what it is signing is contained within the signature, it is not limited to one case (usually the encapsulated message) as most signature protocols are. XML Signature is optimised for XML documents, but it can also be used to sign non-XML documents.

XML Signature makes use of a number of XML standards, including XML Schema to define its precise XML format, and XPath to identify nodes of the XML tree. For transformation purposes it refers to XPointer and XSLT.

Finally XML Signature uses canonicalisation processes to put an XML document into a standard format. An XML Signature can be divided into four parts — the description of what is being signed, the digital signature itself, key information, and other relevant information. The last two parts are optional.

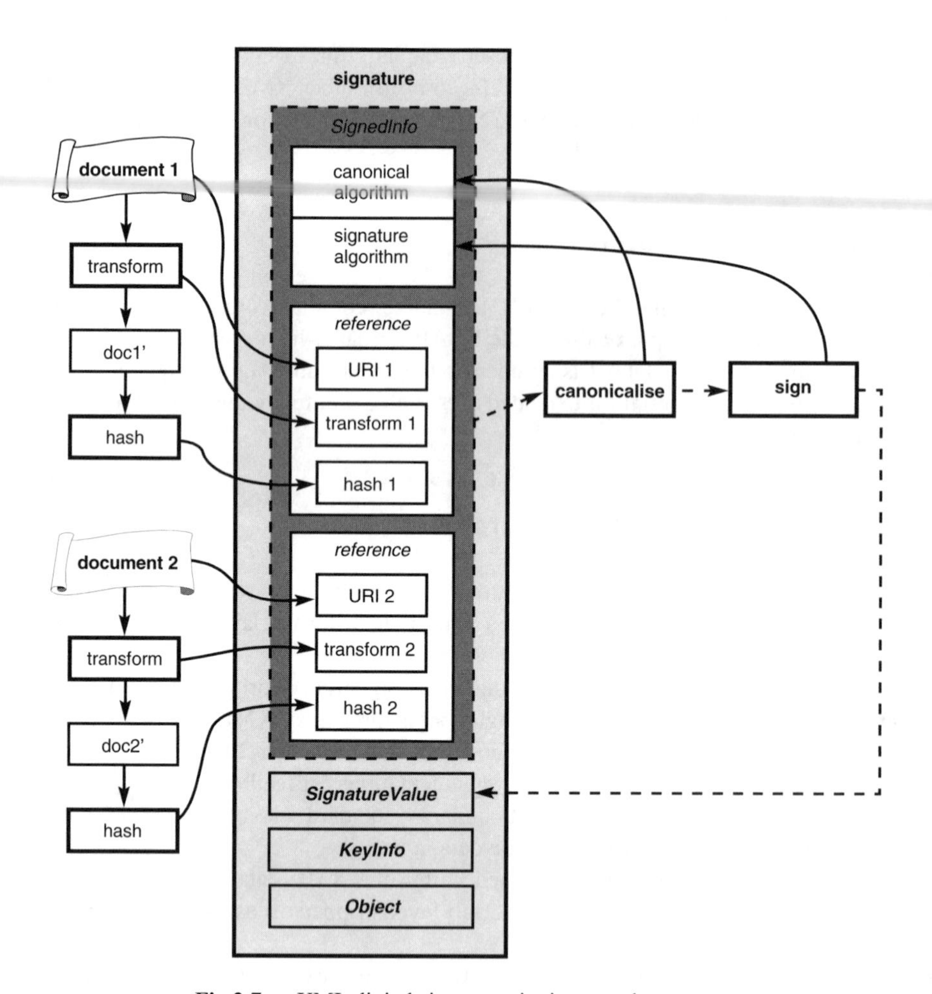

Fig 2.7 XML digital signature signing two documents.

2.6.1 Signed Information

The `SignedInfo` element contains all the details about the signature, including the digital signature algorithm, canonicalisation algorithm and possibly multiple references to the data being protected.

The text contained within the `SignedInfo` tags is what is actually signed, rather than the data references themselves. The `SignatureMethod` child element denotes what algorithm the signature is using — popular choices being RSA, DSA or elliptic curve. `CanonicalisationMethod` indicates the canonicalisation method that the `SignedInfo` XML fragment will undergo before being signed. `Reference` contains the URI-reference and optional transformations that the data at the URI will undergo before being signed. The schema for `SignedInfo` is shown in Fig 2.8.

```
<element name="SignedInfo">
<complexType>
  <sequence>
    <element ref="ds:CanonicalisationMethod"/>
    <element ref="ds:SignatureMethod"/>
    <element ref="ds:Reference" maxOccurs="unbounded"/>
    </sequence>
  <attribute name="Id" type="ID" use="optional"/>
</complexType>
</element>
```

Fig 2.8 Schema for `SignedInfo`.

Figure 2.9 shows the `SignedInfo` part of a digital signature.

```
<SignedInfo>
    <CanonicalisationMethod Algorithm="http://
  www.w3.org/TR/2001/REC-xml-c14n-20010315"/>
   <SignatureMethod Algorithm="http://www.w3.org/
    2000/09/xmldsig#dsa"/>
    <Reference URI="http://www.example.com/xml/Example.xml">
    <Transforms>
      <Transform Algorithm="http://www.w3.org/TR/
      2001/REC-xml-c14n-20010315"/>
    </Transforms>
    <DigestMethod Algorithm="http://www.w3.org/
      2000/09/xmldsig#sha1"/>
    <DigestValue>j6lwx3rvEPO0vKtMup4NbeVu8nk=</DigestValue>
  </Reference>
</SignedInfo>
```

Fig 2.9 Example showing the `SignedInfo` part of a digital signature.

2.6.2 XML Canonicalisation and the Information Set

The aim of canonicalisation is to ensure that XML documents containing the same intrinsic information have the same binary representation, and therefore the same signature. An XML digital signature is intended to be easily **portable**, but an XML document may change its binary representation when being stored and reused. XML data (including a signature) may be stored in tables in a relational database rather than as XML, losing the original XML representation. Putting the information back in the original format would need all the original choices that the XML document used, such as the choice of encoding, the order of the attributes in each element, and so on. (The Unicode standard creates a Universal Character Set (UCS), encompassing most of the world's languages. UCS Transformation Formats UTF-8 and UTF-16 are two possible choices for encoding UCS, using eight and sixteen bits respectively for most of their characters.)

When taken out of a database and put back in XML form, it would be useful if the digital signature still worked. This is the reason for the canonicalisation of XML, which specifies a precise XML format, making clear what to do when there is a choice. This allows the precise binary form of the original digital signature to be recaptured when validating without requiring that the original XML document is retained. Canonical XML is based on information preserved by the popular parsing APIs, such as SAX and DOM.

The Canonical XML standard allows the canonicalisation of well-formed XML. One such case is the order of attributes within an element. The XML standard states that attributes can be created in any order without changing the document meaning. For example the following two elements:

```
<purchaseOrder item="widgets" quantity="3"/>
```

and

```
<purchaseOrder quantity="3" item="widgets"/>
```

contain identical information. Canonical XML determines that attributes should be in alphabetical order, so the first case is the canonical form. Canonical XML is based on the XML Information Set (Infoset), which describes the information contained in a well-formed XML file. XML Infoset shows what information matters and what is irrelevant in transforming the XML file from a raw stream of bytes into data fields and values. For example, the use of UTF-8 or UTF-16 encoding for text is irrelevant, as is the order of attributes. With the XML Infoset, an XML database will know what information from an XML file it has to store and what it can discard without changing the XML document. Canonical XML therefore defines a precise binary stream representation for an XML Infoset.

Validating an XML document using a schema provides far more information than simple well-formed XML. Attributes may have default values, and character data may have an associated type rather than merely being considered as text. XML Schema is defined to create an enhanced information set, called the post-schema-validation (PSV) infoset. This could be used in the future to create an enhanced canonicalisation algorithm for schema-validated XML;

```
<decimal>5.0 </decimal>
```

and

```
<decimal>5</decimal>
```

could be considered equivalent if the simple type is decimal, but not if the simple type is string.

Although the canonical XML standard protects against many changes in binary form of an XML document, there are many that it does not. For example, the two documents in Fig 2.10 have the same meaning, the only difference being namespace prefix. Note that the namespaces themselves are identical (`http://www.example.org/document`).

```
<?xml version="1.0"?>
<a:document xmlns:a="http://www.example.org/document">
    <a:example"/>
</a:document>

<?xml version="1.0"?>
<b:document xmlns:b="http://www.example.org/document">
  <b:example"/>
</b:document>
```

Fig 2.10 Two examples with the same meaning but different namespace prefix.

It is difficult to create the same canonical XML representation in this case, since sometimes data within an attribute or text refers to an element by namespace prefix. Since canonicalisation currently does not require validated XML, it could not know the data within an element is a QName, so could not know it would have to canonicalise this properly. For example, an XML schema fragment (see Fig 2.11) has element references to elements `SignedInfo`, `SignatureValue`, `KeyInfo` and `Object` in the namespace `http://www.w3.org/2000/09/xmldsig#`. [The attribute element also references a QName for its type attribute (in other words the attribute Id has type ID from namespace `http://www.w3.org/TR/2001/XMLSchema`).]

```
<?xml version="1.0"?>
<schema xmlns="http://www.w3.org/TR/2001/XMLSchema"
    xmlns:ds="http://www.w3.org/2000/09/xmldsig#">
[...]
<element name="Signature">
 <complexType>
    <sequence>
      <element ref="ds:SignedInfo"/>
      <element ref="ds:SignatureValue"/>
      <element ref="ds:KeyInfo" minOccurs="0"/>
      <element ref="ds:Object" minOccurs="0"
      maxOccurs="unbounded"/>
    </sequence>
  <attribute name="Id" type="ID" use"optional"/>
  </complexType>
</element>
[...]
</schema>
```

Fig 2.11 Schema fragment with non-standard element references.

Suppose the namespaces were standardised to a1, a2, a3 (lexicographically according to namespace URI) then the XML document would change as shown in Fig 2.12.

```
<?xml version="1.0"?>
<a1:schema xmlns:a1="http://www.w3.org/TR/2001/XMLSchema"
      xmlns:a2="http://www.w3.org/2000/09
      xmldsig#">
[...]
<a1:element name="Signature">
<a1:complexType>
<a1:sequence>
    <a1:element ref="ds:SignedInfo"/>
    <a1:element ref="ds:SignatureValue"/>
    <a1:element ref="ds:KeyInfo" minOccurs="0"/>
   <a1:element ref="ds:Object" minOccurs="0"
     maxOccurs="unbounded"/>
  </a1:sequence>
  <a1:attribute name="Id" type="ID"
use="optional"/>
  </a1:complexType>
</a1:element>
[...]
</a1:schema>
```

Fig 2.12 Fragment of Fig 2.11 with standardised namespaces.

The canonicalisation process cannot change the attribute values `ds:SignedInfo` to `a2:SignedInfo` because it does not know it has to — `ds:SignedInfo` is an attribute value and this could easily be a string rather than a QName value. In other words Canonical-XML does not schema-validate the document to recognise which attributes and elements are QNames — because schema-validated XML is not the only option. The new canonical document now does not make sense to a schema processor, since `ds:SignedInfo` is an element from an unknown namespace. This is the reason that canonical-XML retains namespace prefixes as in the original document. Perhaps a future canonical XML standard for schema-valid XML will be able to canonicalise the namespace.

2.6.3 Referencing Documents

A digital signature refers to the document it is protecting using a URI-reference together with a digest of the data obtained from the URI. A transformation (such as canonicalisation) can be applied to the data before hashing. The use of URI references means that an XML signature signing or verification procedure requires access to a network. The only exception is when the URI-reference refers to the signature document itself. The first case is called a **detached** signature, while a signature that is part of the document it signs is called an **enveloped** signature. The third case, where the reference is part of a manifest of the Object element, is called an **enveloping** signature and is discussed later.

There are subtle yet significant differences between a detached and an enveloped signature that are not initially apparent. Although a detached signature will be a smaller file, it requires on-line validation. It also requires the signed documents to be available on-line, which may not always be the case. However, as a detached signature can contain multiple references, it can be used for a signature over multiple documents.

An enveloped signature can be validated off-line, and there will be no problem with broken links, network outages and other on-line problems. However, an enveloped signature always needs to transform the document to avoid trying to sign the signature value itself.

A signature with multiple references could be both detached (referencing other documents) and enveloped (referencing itself). This could be used for an exchange of messages, where the signature protects the entire message thread up to the current message, so that a signed message cannot be quoted out of context.

2.6.4 Transformation

The transformation could be XPath, XSLT, canonicalisation, a proprietary transformation algorithm, or a combination of transformations. A transformation may be required if the document is signing itself, to remove the actual signature from the data to be signed. Otherwise signing would be impossible since the value of the signature cannot be known before signing (and the signing algorithm needs to know what is being signed). Another case is when a form requires the signature of the sections that should be filled in. However, there are other sections on the form that the form-owner uses for annotation purposes. These should be discarded by the transformation so that they are not to be protected by signature, and can therefore be altered freely.

The problem is understanding what is lost by a transformation. If the transformation T is one to one, then no information is lost in the transformation and the original document cannot be changed. Mathematically one-to-one means $\forall x, y: T(\ ) = T(\ ) \Rightarrow x = y$. Then, knowing the hash digest from the signature information, there can only be one transformed document $T(x)$ which means that there can only be one original document by the one-to-one rule. There is also the conceptual problem of knowing what is being signed, the document or the transformed document. The original document is represented by the URI-reference, the transformed document by the transformation and the hash digest — so a case can be made for both.

2.6.5 Signature Value

The `SignatureValue` element contains the actual value of the signature, base-64 encoded. The actual data that is signed is the canonicalised `SignedInfo` element and its children. Since this contains the hash value of the URLs, the documents at the URL are protected.

2.6.6 Key Information

The `KeyInfo` element is optional, and allows the inclusion of key material with the digital signature. A number of methods for finding the relevant public key for validating the signature is specified. The problem with asymmetric cryptography has always been connecting a public key to its owner. The most popular current mechanisms for achieving this are PKI, PGP and SPKI, and these are incorporated within XML Signature. The signer could also specify their own methods by extending the specification.

The most basic key information is the public key itself, and this may be acceptable if there are alternative means of enabling trust, or for test cases. The second possibility is providing a public-key certificate (base-64 encoded). XML Signature defines three different certificate types — X.509 certificates, PGP key identifier, or an SPKI certificate.

Retrieving the key from another location using a URI-reference is supported using the `RetrievalMethod` element. This is also useful if there are multiple signatures in one document, allowing the public key information to be specified only once, and referred to by the other signatures. A key can also be referred to by name. This may be useful in small closed groups, where everyone's key is manually known.

The XKMS specification makes use of the XML key information elements for key registration and recovery. For registration, an XKMS service may take a key value and return a key name or certificate. For recovery it may take a key location or key name and return a key value (Fig 2.13).

2.6.7 Additional Information (the Object Element)

`Object` is an optional element that allows an XML signature to contain additional information. It allows any XML data — so a timestamp could be put here to indicate the time of signature. XML Signature suggests two structures that an object could use. The `Manifest` element is a collection of URI references and digests similar to the `Reference` element in the signed information section. Putting the references in a manifest conveys a different validation meaning — that each reference does not need to be checked against its digest. Such a signature is called an enveloping

```
<complexType name="KeyInfoType" mixed="true">
<!--The KeyInfo type -->
  <choice maxOccurs="unbounded">
    <element ref="ds:KeyName"/>
      <!-- Name of a key -->
    <element ref="ds:KeyValue"/>
      <!-- Actual public key value -->
    <element ref="ds:RetrievalMethod"/>
      <!-- How to retrieve the key -->
    <element ref="ds:X509Data"/>
      <!-- An X.509 certificate -->
    <element ref="ds:PGPData"/>
      <!-- A PGP public key identifier -->
    <element ref="ds:SPKIData"/>
      <!-- A SPKI certificate or key pair -->
    <element ref="ds:MgmtData"/>
      <!-- In-band key distribution data -->
      <any processContents="lax"
       namespace="##other"/>
    <!-- Other methods -->
  </choice>
</complexType>
```

Fig 2.13 Example of XML key information elements.

signature. The second structure is a collection of signature properties. This is where timestamps and information such as time to sign could be placed. It could even be the place to put information to help the de-referencing of a URI.

2.6.8 The Validation Process

Validation is split into three parts. Firstly, the signer's public key must be obtained by some means. This may be obtained using the key information provided by the XML signature. However, in some cases the public key may already be known or it can be obtained by some other mechanism. Secondly, the signature value can be checked by processing the `SignInfo` element according to the stated canonicalisation and hash algorithms to obtain a hash value and comparing the hash against the signature using the public key as in a normal signature verification. This is called **signature validation**. Finally the URI references must be checked for the correct digest value. This part is called **reference validation**. XML Signature specifies that all these steps must be fulfilled before the signature is considered valid. Verification failure would result if one of the references fails to provide the correct digest, or the reference cannot be resolved due to connection difficulties. If a signer does not wish to make reference validation compulsory, the references can be put in a `Manifest` element. This is called an enveloping signature.

There may be problems with any digital signature legislation around the XML signature standard. For example, how can it be decided legally whether (in the detached signature case) reference validation was successful at a certain time in the past? A party may wish to cover up responsibility by stating that the XML document to which the signature referred was never present at the URI stated by the signature and consequently that the signature was never valid. Suppose that it was removed at a certain date but remained in a cache — does the party still have responsibility for a signature validated after this date?

There are other potential problems with using external references. A URI could require either a cookie or authentication in order to be accessed. An example scenario is when a policy recommendation is made based on information at an external Web site, and a digital signature is employed referencing the external Web page to ensure that the recommendation is only valid if the external information is not changed. This information is copyrighted, and therefore cannot be placed in the document itself. Consequently a detached signature is the only valid method. However, if a cookie is needed to obtain access to the Web site, it is hard to create the digital signature and allow the reference validation to work properly. A solution could be to place the cookie information within a `SignatureProperty` element of the `Object` element. This step, however, is not part of the defined reference validation process, so an XML Signature implementation could correctly declare the signature invalid due to failure to access the URL.

2.7 Problems with XML Signature

XML Signature is a generic standard, covering most ways by which a digital signature in XML format could be employed. In most cases a digital signature used in a protocol or legal document needs to be narrowly defined to ensure that it cannot be abused. A protocol may require that the signature needs to be enveloped only. The allowed transformations may be restricted to a few cases. Key information may be required in one particular format. Therefore it is likely that the XML signature elements will be redefined in many cases, restricting its usage. Hopefully there will be standard ways to do this, so that each protocol or standard, making use of XML Signature, does not have to continually ‘reinvent the wheel’.

2.8 Uses for XML Signature

2.8.1 Digital Signature of Web Page

Signing a Web page will allow humans to understand that information conveyed by it is authentic. A typical example is a historical document that has been displayed as an XHTML Web page and is signed by an archivist or historian to indicate that it is genuine. Then it can be quoted or referenced by historians in their own writings. This is better than obtaining the document using a secure link to an authenticated Web server because it is the document itself, not the Web server (or the connection), that is required to be genuine.

In addition, the document can be placed at a different location and still retain the security of the digital signature.

2.8.2 Security Protocol Using XML Signature

Suppose there is an authentication protocol which results in the authenticated client receiving a token at the end of the process; this authentication token can then be used to verify these credentials to other parties on the network. If the token uses an XML format, it can contain data created using the multitude of XML vocabularies that are available. The authentication server's security approval will be indicated by the XML signature of the XML data within the token.

2.8.3 Machine Interpretable Trust Mechanisms

If the XML Signature standard can be combined with a basic logic-processing language, there is the basis for trust mechanisms where decisions are performed by machines based on the logical rules set up by the owner, or by a client where the trust decisions have been laid out logically by the client machine. For example, a decision to download code from an unknown company is a potential security threat. The signed code and the public key certificate connect code to company. The client machine could then gather signed statements on the business from accountants, auditors, customers, and business analysts to determine the validity, reputation and financial health of the business. These actions could be performed using Web services. The accumulated information could then be presented to the client in order to make the final decision.

2.9 XML Encryption

Recently the W3C have set up an XML Encryption activity, aiming to provide a standard for encrypted information within an XML document. This could be used to keep confidential parts of an XML document. For example, there could be secret clauses in a contract that only certain parties are allowed to know.

XML Encryption could be employed for alternative non-signature integrity mechanisms such as a keyed hash function. Web services which contain confidential data in certain fields may use XML Encryption.

The XML Encryption standard will define symmetric key and associated data types for use within XML. XML Encryption is required to support the encryption of non-XML resources, and therefore could be used to protect a future pay-per-view Web broadcast, with scrambled data unlocked at a later date on receipt of payment. XML Encryption may use transformations in a manner similar to XML Signature. However, there could be different reasons for using them. Firstly, a transformation may help improve the security by adding redundancy. It could be a compression algorithm to again help security and also shorten the document length.

The uses for XML Encryption are not so immediately obvious as for XML Signature. A possible use will be where XML is being employed to serve as layers in a network protocol. Routing layers may need to be in plain text, but the actual message needs to be encrypted. The example in Fig 2.14 is from the Jabber [15] instant messaging protocol. Since routing is determined by the Jabber server and is not directly client-to-client, the encryption needs to exclude this information so that the server can read it.

```
<message from="juliet@capulet.com/balcony"
    to="romeo@montague.com/orchard">
  <body>This message is encrypted</body>
  <x xmlns="jabber:x:encrypted">hQE0A7wG+dhHej/
    Aheif8s5G8UesGh[...]</x>
</message>
```

Fig 2.14 Example of Jabber instant messaging protocol.

XML Encryption could be used in Web pages when there are no alternative authentication or access control mechanisms, and an end-to-end encryption scheme is not possible. However, in most cases end-to-end encryption is more desirable, so mechanisms such as SSL and IPsec will be more suitable. Encryption cannot be employed in a situation analagous to the detached signature, and there is little reason for portability. It may have use in key encryption schemes, to provide the key for an end-to-end encryption scheme. For that reason, key-encryption Web services may be a future possibility. It also could be used to encrypt XML (and other) files on a computer.

There are several security issues that an XML encryption standard will need to address. Firstly, if the amount of data being encrypted is small, then it could be guessed from the structure of the XML. Secondly, information may be leaked by a schema. Other problems include changing the data type of an element, attribute or character data. If schemas are employed, this may change the schema. Therefore encrypted XML may need to employ two schemas, one for its encrypted state, and the other for its decrypted state.

2.10 Summary

XML Signature and XML Encryption provide the basis for a new type of security mechanism, and for enabling secure Web services. The advantages of XML Signature over other digital signature standards are its flexibility and its portability. Both standards bring the security 'vocabulary' to XML documents, allowing agreements on levels of security to be made using XML. With XML Signature and XML Encryption, security mechanisms can be interwoven within a document, interacting and making use of its data.

References

1 ISO — http://www.iso.ch/

2 Internet Engineering Task Force — http://www.ietf.org/

3 XML Signature standard — http://www.w3.org/TR/xmldsig-core/; XML Signature is also an IETF draft standard — http://www.ietf.org/internet-drafts/draft-ietf-xmldsig-core-2-00.txt

4 Early proposal to standardise HTML using a layered structure — http://www.w3.org/MarkUp/HTML-WG/940726-minutes.html

5 Official XML standard — http://www.w3.org/TR/REC-xml

6 DOM level one — http://www.w3.org/TR/REC-DOM-Level-1/

7 XML-Data (W3C note) — http://www.w3.org/TR/1998/NOTE-XML-data-0105/

8 W3C note: '*Document Definition Markup Language*' (DDML) — http://www.w3.org/TR/NOTE-ddml

9 Schema for Object-Oriented XML (version 2.0) — http://www.w3.org/TR/NOTE-SOX/

10 A few XML introductory links are — http://www.xml.com/pub/a/98/10/guide0.html, http://www.w3.org/XML/Activity, http://www.w3.org/XML/1999/XML-in-10-points (see Ref 5 for the definitive treatment).

11 Meatxml — http://www.meatxml.org

12 XML Namespace standard — http://www.w3.org/TR/REC-xml-names/

13 XML Schema Working Group — http://www.w3.org/XML/Schema

14 SAX — http://www.megginson.com/SAX/

15 Jabber Technology Group — http://www.jabber.org/

3

USING XML SECURITY MECHANISMS

A Selkirk

3.1 Introduction

Now that the security standard XML Signature has completed its standards track and XML Encryption is following close behind, the security protocols and software to use them can be created. XML Signature is a sophisticated, generic standard for creating digital signatures in an XML format, optimised for signing XML data. As there are many business formats and standards that already use XML for representing data, there is a ready market in which XML Signature can be applied. However, the major commercial use may be to secure Web services. There are already a number of proposed security service standards that aim to fill this niche. This chapter explores how new security mechanisms will be employed for new Web technologies. It contains an investigation of Web services, showing how XML is being used for the next generation of distributed architectures and eCommerce projects. The current building blocks of a public key infrastructure are analysed with an intention to see whether XML can create improvements. The new XML security standards, such as the security assertion markup language and XML Key Management Specification, show how XML Signature and XML Encryption can be used to create sophisticated new security services and protocols. This chapter aims to determine how these new mechanisms could operate with familiar protocols, such as the secure sockets layer, to create an overall security infrastructure, usable in an Internet environment.

3.2 Web Services

The latest buzzword in the computer industry is Web services, a collection of technological ideas and standards that every major software company (e.g. Microsoft, IBM, Sun, Oracle) is working on. A service can be defined as a computer application that provides computational or informational resources on request. A service that is accessible by means of messages sent using standard Web protocols, such as the hypertext transfer protocol (HTTP), is called a Web service. The intention is to give businesses mechanisms for broadcasting their services and agreeing to contracts — in effect a front-end virtual shop window for their services (see Fig 3.1). Web services may communicate with other Web services in order to acquire necessary information, or fulfil parts of a contract. Therefore they have a distributed computing architecture. They use XML to describe the information exchanged between vendor, buyer and third parties. They work over existing Internet protocols such as TCP/IP, SMTP and HTTP, but with an XML payload.

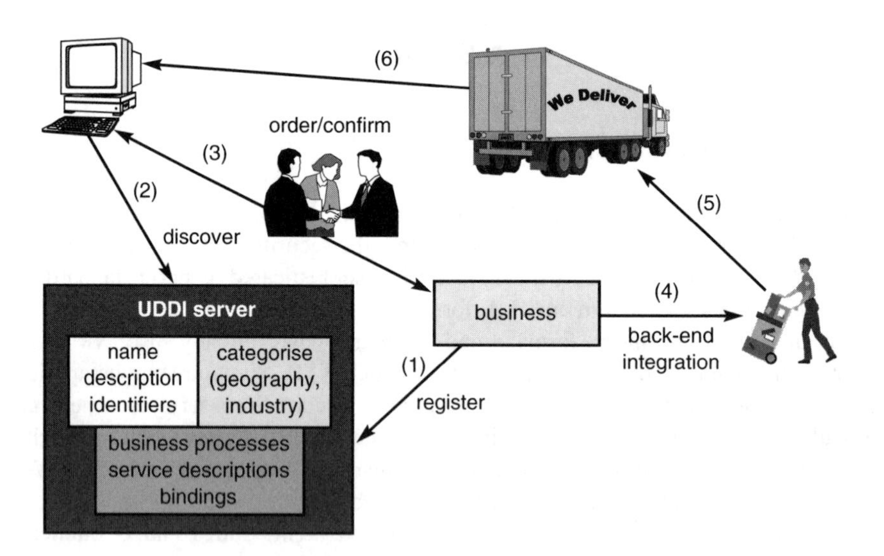

Fig 3.1 A Web service transaction, using UDDI and SOAP.

TCP/IP sockets have already enabled a number of service protocols. The most well-known are SMTP for e-mail, FTP for file transfer, NNTP for newsgroups and HTTP/HTML for Web display. There are many other socket-based protocols that have been defined by the IETF and other groups. Some have been successful, but others have never been used on a large scale. Basing Web services on XML and HTTP (or SMTP) does not guarantee success where older, similar protocols failed. XML gives operating system and language independence, and easier comprehension (a binary standard such as ASN.1, used in the X.509 standard to define a certificate format, is not easy to work with). HTTP provides basic error messages and is already widely used. None of these is such a definite advantage that Web services are bound to achieve universal use. However, the support for Web services in the software industry may create its own momentum regardless of technological advantage.

3.3 Web Service Protocols

The major components of Web services are SOAP (the Simple Object Access Protocol) [1], UDDI (Universal Description Discovery and Integration Specification) [2], WSDL (Web Services Description Language) [3], XAML (the Transaction Authority Markup Language) [4] and ebXML (eBusiness XML) [5], all of which are XML formats.

3.3.1 Simple Object Access Protocol

SOAP provides a model for simple two-way XML communication, which could become the leading distributed applications technology. It aims to achieve success in the open Internet environment where CORBA, DCOM and other distributed object technologies largely failed.

This does not necessarily mean replacing DCOM and CORBA where they work successfully, mostly in intranet-type environments. SOAP uses XML as a platform-neutral format to transfer data and is defined for use over HTTP and SMTP, allowing leverage of existing technologies. The use of the hypertext transfer protocol is well thought out, since a SOAP server will just be a Web server with a module to handle SOAP calls.

Figure 3.2 shows a typical SOAP request message, using HTTP. A client is asking a SOAP server for the latest trade price of the company with stock symbol 'DIS'.

```
POST/StockQuote HTTP/1.1
Host: www.stockquoteserver.com
Content-Type: text/xml; charset="UTF-8"
Content-Length: nnnn
SOAPAction: "Some-URI"
<soap:Envelope
     xmlns:soap="http://schemas.xmlsoap.org/soap/envelope/"
     soap:encodingStyle="http://schemas.
     xmlsoap.org/soap/encoding/">
   <soap:Body>
    <m:GetLastTradePrice xmlns:m="Some-URI">
     <m:Symbol>DIS</m:Symbol>
    </m:GetLastTradePrice>
   </soap:Body>
</soap:Envelope>
```

Fig 3.2 Typical SOAP request message.

The SOAP response message (Fig 3.3) is also a relatively straightforward affair.

```
HTTP/1.1 200 OK
Content-Type: text/xml; charset="UTF-8"
Content-Length: nnnn

<soap:Envelope
      xmlns:soap="http://schemas.xmlsoap.org/soap/envelope/"
      soap:encodingStyle="http://schemas.
      xmlsoap.org/soap/encoding/">
   <soap:Body>
      <m:GetLastTradePriceResponse xmlns:m="Some-URI">
      <m:Price>34.5</m:Price>
     </m:GetLastTradePriceResponse>
   </soap:Body>
</soap:Envelope>
```

Fig 3.3 Typical SOAP response message.

Figures 3.2 and 3.3 show SOAP used in a distributed computing environment. Although the communication is basic server-client 'question and answer', there are projects to extend this to provide the type of reliable communication found in traditional distributed computing architectures. One such additional necessity is security. SOAP has been submitted to the W3C and will be one of the inputs to the new W3C XML Protocol working group.

3.3.2 Web Services Description Language

WSDL allows a service to describe its data format for communications — the inputs a service takes and the outputs it provides. WSDL also specifies the protocol over which a service runs (e.g. HTTP, SMTP), and other necessary information. Computers passing SOAP messages can thereby understand both the data format required to send a particular message and the structure of what will be received. In other words WSDL connects SOAP request and response messages, using XML schema types to describe the XML data.

3.3.3 Universal Description, Discovery and Integration

UDDI is an initiative to allow businesses to register Web services and for customers to find them. It currently has three information models — white pages for business names and descriptions, yellow pages for categorisation by industry and geography, and green pages for business processes, service descriptions and bindings. The registry is distributed over 'operator' sites which share information on a daily basis

— a model similar to DNS servers that handle the TCP/IP domain name registry. DNS was a major contributor to the success of the Internet and World Wide Web by providing a mapping from English words to an unmemorable Web address — a contribution shared with Web search engines that can aid in finding Web sites and information. UDDI could prove to be of similar importance for business-to-business eCommerce.

3.3.4 Transaction Authority Markup Language

For business-to-business Web services to work, transactions need to be part of the model. Ensuring that every part of a complex multi-party deal either succeeds or fails together is not a trivial task. XAML aims to create this functionality. Transactions are a basic part of database technology, but ensuring transactions succeed across multiple databases is more difficult. With such an API in an XML format, on-line Web services can become real multi-party business processes.

3.3.5 The Current Situation for Web Services

Web services will require other technological pieces, such as the back-end business integration of Web services to databases and legacy systems. The current set of standards is not uniform, there being competing and overlapping standards, although this is not yet a serious problem. There are also moves to take Web services further and create a model for leasing software as a service over the Internet. An example of this is providing key management services to support the signing and encryption of messages.

3.4 Public Key Infrastructure in a Web Service Environment

Generally the security services required by client/server applications are authentication, access control, confidentiality, integrity and non-repudiation. Authentication and access control separate users who are allowed to perform certain actions from those who are not. Confidentiality ensures that conversations and messages are kept secret between sender and recipient, while integrity ensures that messages cannot be altered without being noticed. Non-repudiation services aim to keep parties honest by preventing the denial of an action. These may all be required by Web services to prevent fraud, abuse of a service, or actions not permitted to a person.

Web services are targeted for use in an open network environment. If anyone can potentially access the Web service, some authorisation or access control will be needed unless all Web service actions are permitted to everyone. The general method of distinguishing different entities is by binding an entity to its public key and using protocols to establish the ownership of the private key. This requires a public key infrastructure (PKI) of some form. There are a number of different models for PKIs, and they have a lot of different parts. There are many PKI services that have to be performed by on-line entities, from keeping a list of trusted chains in the PGP model to providing key revocation in a conventional certification authority model.

These can be created as Web services in their own right — although it must be remembered that there is no advantage in changing a perfectly good existing protocol, that is already widely used, into a Web service for its own sake. The following sections examine the parts of a PKI, considering their suitability in a Web service environment.

3.4.1 Certification Authority

A certification authority (CA) is an organisation that binds entities to their public keys. The idea is that a CA has its own public key. When satisfied that Alice owns a public key, the CA creates a signed document that contains information about Alice (usually Alice's name), information about the CA, and the value of Alice's public key.

This signed document is called a public key certificate. The certificate can be validated by others using the CA's public key. Therefore, on assuming the CA is competent and trustworthy, the certificate can be used to prove that information verified using the public key contained in the certificate was originally signed by Alice.

A certification authority will issue many public key certificates for lots of entities. This reduces the problem of being able to securely obtain the public keys for any of these entities to knowing the public key for the certification authority and being able to trust it.

If a certification authority is sufficiently large and well-known, it could publicly broadcast its public key in a way that cannot be replicated by an imitator. Alternatively the CA could obtain a public-key certificate from another larger CA. This is known as a pyramid of trust. As an example, the BT TrustWise [6] CA has a public key certificate obtained from the VeriSign [7] CA.

A certification authority may provide a number of services for its users, including certificate registration, revocation and certificate directories. Some of these can be performed on-line, while tasks such as the verification of credentials may only be possible off-line.

Even here some credential checking could be carried out on-line — the verification of an e-mail address for example. In a Web service world, many of the on-line services could become Web services.

3.4.2 Local Registration Authority

The local registration authority (LRA) model is a notable variant on the CA model and is used by many large organisations. Identity verification and trust is performed locally by the LRA, but the certificate is issued by the CA. This allows the large organisations to determine who their employees are, but the certificates will be acceptable to anyone who trusts the CA. Most certification authorities sell such agreements — for example the VeriSign OnSite service [7].

There are at least two protocols for CA/LRA communication — the certificate request syntax (CRS) and the Internet X.509 public key infrastructure certificate management protocols (PKIXCMP). If more complex XML certificates are required, communication could be recast as a Web service.

3.4.3 Certificates

The most commonly used certificate today is the X.509 (version 3) public key certificate. Certificates can be used in many different situations and for lots of purposes. Software companies can use them to protect the integrity of their applications and to guarantee their security. Businessmen may use digital signatures to complete a deal on behalf of their employers. Digital certificates can be used in authentication exchanges to allow access to different sites. X.509 v3 certificates have an extension field which can be used to indicate the manner in which the certificate should be used. Unfortunately there is no universally defined structure that can be put there and understood by all. For this reason it is possible that there will be future certificate standards in an XML format. This will allow flexibility in the data put in the certificate. A certificate could contain statements limiting how it should be used. For example, a certificate could be limited to verifying downloaded data, or it could be used in an authentication process, or to allow secure financial transactions.

3.4.4 Directory of Certificates

In the early days of asymmetric cryptography and public key infrastructure work, there was an attempt to create a universal directory for public keys. This was based on the concept of telephone directories, which linked identity to telephone number. The work produced the X.500 set of standards, but failed to achieve universal directories due to political and technical barriers. A number of technologies came out of the failed process, including the X.509 public key certificate and the lightweight directory access protocol (LDAP) [8], a scaled-down X.500 directory. Public key certificates are a useful tool for digital signature and encryption, but obtaining them can be a problem. In the case of digital signatures, a signature can always be accompanied by a certificate, although this is a little wasteful in bandwidth if the certificate is already known. For public-key encryption, a person wishing to send a message has to look for the correct public key with which to encrypt. This is where a directory of certificates is useful. Web services that provide this functionality may be viable.

The problem of finding the correct directory service raises the question of a universal directory again, and this may again be attempted for certain types of use. The UDDI standard has ambitions for universality in business data, but there is a recognition that this is not an easy task. Signatures and encryption have a wide range of uses, not all requiring such services. However, a universal security mechanism, such as one employed over the World Wide Web may require a universal directory.

3.4.5 Registration for Certificates

There are many processes for applying to a certification authority for a certificate. The cheapest certificate (class one) is for personal use, and generally just links a public key to an e-mail address. Such a certificate can be bought on-line using a credit card and supplying an e-mail address. Higher grade certificates require better verification checks, such as face-to-face contact, checking passports, and double-checking any information that the certificate may contain. A large business can register its own employees by becoming a local registration authority through agreement with a certification authority. Registration mechanisms could be as a Web service using SMTP rather than HTTP to take into account delays while physical verification is performed.

3.4.6 Revocation

There are two parts to a revocation. Firstly a certification authority must revoke a certificate issued by it. Then each individual verifying signature must obtain notification of the revoked certificate to ensure that signatures signed using the public key are no longer trusted.

Revocation is always a major problem. The situation where a revocation list is checked every time a signature is validated would be a slow process, not scalable, and not much use in an off-line environment. It also negates most of the benefits of PKI checking a secure directory for symmetric keys. Revocation could be managed as a Web service. It could also be checked once every time period. It could be negated by having short-term certificates. Generally it should be the policy to restrict long-term certificates to unchanging or low-security attributes, such as name. This will limit the necessity to revoke certificates due to incorrect information, but there will always need to be revocation due to private key compromise.

3.4.7 Short-Term Certificates

Short-term certificates are usually issued for a short-term purpose or in the case where a person acquires an attribute for a short period of time. For example, an employee uses a long-term identity certificate as credentials to obtain a short-term 'entitlement' certificate from a company certificate server that allows the employee to carry out purchases on behalf of their company for that day.

3.4.8 Attribute Certificates

Attribute certificates are certificates that are issued based on a particular use, rather than tying identity to public key. They are generally short-term, thus allowing for changes in role or security policy. XML is particularly useful for attribute certificates because XML standards can provide the vocabularies for attributes. This creates the possibility that attribute certificates could become portable across security domains. The requirements are that the attribute certificate server is trusted to certify certain attributes by both domains, and that the XML languages used are understood by both. For example, a bank could issue attribute certificates to client businesses, stating that their financial position is good. This could be used by the businesses when bargaining for contracts. This attribute certificate is necessarily short-term, since financial positions change on a day-by-day basis. There are many IETF draft standards on the format of attribute certificates [9], but as yet there has been no standard that has achieved large-scale success. An XML standard such as the security assertion markup language could be the key to popularising short-term authorisation tokens.

3.4.9 SPKI and PGP Models

There are alternative models to conventional CA-driven identity certificates. These can work easily within XML security models, since XML security mechanisms are generally certificate-type neutral. If the actual models are converted to an XML format, they may become more flexible. For example, PGP works on chains of trust between individuals, but, using XML, each chain could also contain reasons for the trust relationship. XML could help in refining the trust model. Often trust is related to certain actions that an individual can perform. For example, in knowledge-based decisions, a person may be trusted to be competent in one area of expertise but not another. Therefore XML could be used to provide PGP-type trusted links, but only on certain actions, with XML standards providing the vocabulary to describe the actions.

3.4.10 Trust Choices

In the end, there is always a choice between accepting a public key for a particular purpose, or rejecting it. This decision could be helped by a variety of Web services that could be used to provide information on which to base this decision. There are questions of identity, authority, liability and risk. Providing mechanisms to retrieve this information, and bringing it all together so that the decision can be made, should be part of the security infrastructure.

3.5 New Security Problems

The use of new computing models such as Web services and the increased sophistication of Internet-related software means that there are new security problems that need to be tackled. In addition, there are issues that have not been satisfactorily solved yet, such as single sign-on.

3.5.1 Utility of Firewalls

A firewall controls access between two separate networks. It is generally used to control information flow to and from a corporate intranet to the Internet. A firewall is designed to block hostile threats to internal computers, while allowing general Web access features to intranet users. The increasing use of XML provides new problems to firewall designers. Since code, macros and other potentially dangerous functionality can be embedded within XML, it becomes increasingly difficult for firewalls to separate what is dangerous from what is not. SOAP runs over the hypertext transfer protocol, and so it uses port 80 just like any normal Web traffic. This makes distributed application communication such as client-to-server or peer-to-peer resemble normal Web-page retrieval. It would need a sophisticated analysis of normal Web traffic to separate legitimate from non-legitimate use of Web services. However, even the best firewalls can only hope to block certain actions that are not desired, not all undesirable actions — unless all traffic is prevented, which negates the purpose of linking networks together. There is a model that sees XML processing involving a series of Unix-like processing pipes — XML information can be sent to applications interested in that particular piece. A future firewall could use this concept to monitor (and possibly alter) XML information in which it is interested. However, this involves far too much processing for a firewall to achieve. The answer may be to enforce security policy at the individual computer level — in the operating system, or perhaps in hardware at the interface to the network.

3.5.2 Limitation of Certificate Usage

When a certification authority issues a public key certificate to an individual, the granularity is not high. For questions of identity, especially when sending encrypted messages to the individual, this does not matter. However, when the private key is used in authentication mechanisms, or to sign code for download, the simple X.509 certificate may not indicate whether these actions are appropriate for the certificate owner. Each action requires a different level of trust. What is required are certificates tying private key to actions. As separation of use is a vital concept in traditional security, restraining an individual's use of a signing key may therefore be good practice for information security.

3.5.3 Portability

There have been many attempts at creating single sign-on authentication environments, to varying degrees of success. One of the largest is the Kerberos [10] security architecture implemented in Windows 2000 and other security systems. The idea of most architectures is that a user, once authenticated, is granted a (signed) ticket. This ticket contains the user's authentication status and entitlements. It can be passed to other similar environments to create the single sign-on requirement. A problem has always been the flexibility of the vocabulary — some authentication systems need one vocabulary, while another system will require something else.

3.5.4 Logical Language for Trust

Often a trust decision can depend on a multitude of factors — information, trusted third parties, degrees of trust, types of decision required, risk assessment, and so on. There is currently no language for trust assertions, or logic processing.

3.6 XML Security Services

A key issue in creating a security architecture is the placement of security within the architecture. Within a network protocol stack, such as the World Wide Web uses, security mechanisms can be placed at many different levels (see Fig 3.4). Generally security placed at a lower layer means a better end-to-end security solution, since all the information in the layers above is protected. The higher the level, the easier it is to combine the security data with the application and the finer granulation of security can be achieved. Generally security at the IP layer (IPsec) can prevent denial of service attacks and create virtual private networks. Security at the socket layer using SSL can ensure the confidentiality of data. It can also provide authentication for both server and client using the Diffie-Hellman protocol. Together, SSL authentication and encryption can provide a non-portable data integrity. The intention for XML signature is to create non-repudiation, integrity, authentication and access control services.

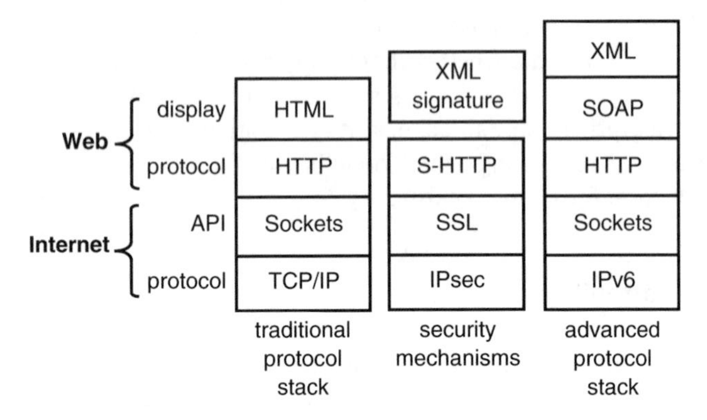

Fig 3.4 The various layers of protocol that a typical Web application uses.

The technological advantages in using XML Signature are granularity, portability, and the cross-utilisation of other XML standards. Granularity comes from being able to pinpoint the part of the XML document to sign or encrypt. It is useful in situations where end-to-end encryption may break a security policy (for example, encrypted e-mail could prevent the elimination of messages containing viruses at the e-mail account server level) or where only a particular part of a document is relevant to the security mechanism. The use of other XML standards can give specific meaning to an XML signature. This is demonstrated in the security assertion markup language (SAML) standard, where the digital signature provides authorisation for restricted actions.

A portable security mechanism is one where the results can be stored and reused in different protocols. Specifically portable security services should be employed across different security domains. Since XML is always the top layer of a protocol stack, XML security services are far more portable than lower level mechanisms. As long as domains understand and can process the necessary XML vocabularies, they can make use of the same XML security mechanisms. For example, an XML signature, being part of the application data structure, is retained by default. By contrast security data structures built into the sockets' API or the IP headers are generally discarded before being processed. Portable security mechanisms are ideal for Web services because they can be used in a chain of Web services. Also the mechanisms can be provided as a separate Web service in their own right.

In November 2000, the first XML specifications to define trust services were published. OASIS, the Organisation for the Advancement of Structured Information Standards [11], has taken charge in producing standards for XML security services. Currently OASIS is working on two specifications:

- SAML (security assertion markup language) creates portable authentication and authorisation information — this could enable single sign-on and authorisation models that cross domain boundaries;
- XACML (extensible access control markup language) aims to standardise access control information.

VeriSign, Microsoft and WebMethods have created XKMS, the XML Key Management Specification. This describes how to provide key management as a Web service, especially in managing keys used with XML Signature. VeriSign have also created X-TASS (XML Trust Assertion Service Specification), an architecture and retrieval protocol for trust assertions. In addition, XML Signature is used directly in a number of business standards, such as ebXML. The following sections describe the major emerging standards. Many of them are in their early stage, and so their final form may be somewhat different.

3.7 XML Key Management Specification

The first security Web service proposal is the XML Key Management Specification (XKMS), created by Microsoft, VeriSign and WebMethods. It has been submitted to the W3C [12], and may be the basis for further work after the digital signature and digital encryption standards have been completed. It intends to convert key management of both asymmetric and symmetric keys. A Web service implementing XKMS is intended to perform key management services on behalf of its clients. There are two parts to the standard — a key registration service (X-KRSS) and a key retrieval service (X-KISS) (see Figs 3.5 and 3.6). Both make use of the key information types in the XML Signature standard as the vocabulary to express data items such as keys, key names, key locations and certificates. For key registration, clients send the XKMS service their public key and relevant identity information and receive a certificate, or a location from which the public key can be retrieved by others.

Registration can optionally involve a protocol by which clients prove ownership of the corresponding private key. Once the key has been registered, a client can utilise further services, such as key recovery (if the private key is escrowed) or key revocation.

```
<?xml version="1.0"?>
<Locate xmlns="http://www.xkms.org/schema/xsms-
2001-01-20" xmlns:dsig="http://www.w3.org/2000/09/xmldsig#">
   <Query>
    <ds:KeyInfo>
      <ds:KeyName>John Regnault</ds:KeyName>
   </ds:KeyInfo>
   </Query>
   <Respond>
    <string>KeyName</string>
    <string>KeyValue</string>
   </Respond>
</Locate>
```

Fig 3.5 An XKMS request using X-KISS for retrieving a public key.

```
<?xml version="1.0"?>
<LocateResult xmlns="http://www.xkms.org/schema/xsms-2001-01-20"
    xmlns:dsig="http://www.w3.org2000/09/xmldsig#">
   <Result>Success</Result>
   <Answer>
    <ds:KeyInfo>
     <ds:KeyName>John Regnault</ds:KeyName>
     <ds:KeyValue>[...]</ds:KeyValue>
    </ds:KeyInfo>
   </Answer>
</LocateResult>
```

Fig 3.6 An XKMS response.

The key retrieval service will be required by a client that cannot retrieve a public key on its own, or cannot determine whether the public key belongs to a trusted entity. This could be used in a configuration where keys are referred to by key name, as in the examples in Figs 3.5 and 3.6. The XKMS service would act as a central repository storing all key names and associated key values. The client then queries the service for the key value on being sent the key name. This allows simple applications with a small trusted community of, for example, a hundred people to have a secure system without requiring the complexity of a public key infrastructure. It is probable that key retrieval mechanisms will be required more for public-key encryption than for digital signatures, since a public key needs to be looked up before a confidential message can be created.

The rationale behind XKMS is to support the use of the XML Signature and prospective XML Encryption standards, to support Web services with key management facilities, and to be a Web service in its own right. The idea is not to sell XKMS servers to organisations, but to provide the facility as an outsourced service. Whether this model will succeed has yet to be determined.

3.8 Security Assertion Markup Language

In late 2000, two XML security specifications were published using XML for authentication information. These were the Security Services Markup Language [13] (S2ML) and AuthXML [14]. Rather than having rival standards, both groups decided to produce a common standard using the OASIS standards organisation. The new standard is called the Security Assertion Markup Language (SAML).

SAML aims to create a language and protocol for using signed assertions, or tokens, to state a user's entitlements. It does not aim to create new authentication protocols, but rather a method of picking up authentication data from existing protocols (such as simple password or more sophisticated authentication). There are several services that could use SAML. Single sign-on is the concept of ensuring that the client need authenticate himself only once, and then using the created authenticated token for access to all other secure services. An authorisation service is one that can delegate responsibility to clients by handing out signed entitlement tokens, usually valid for a short period. In each case a token is created that provides security assertions readable by other parties. XML ensures both portability and the vocabularies necessary for machine-readable security assertions.

As XML Signature is part of a security assertion, technological aspects of digital signatures need to be considered. For example, the validation of a signature generally requires the retrieval of key information, checking revocation lists and verifying the signature. A revocation list is generally used for negating compromised private keys. However, where security assertions are long term, a revocation mechanism for these assertions (rather than the key itself) needs to be part of the infrastructure. In other words, the private key is still perfectly good, it is just that the assertions made for it are not.

Since the use of authentication mechanisms is a major aspect of XML Signature's appeal, there is a lot of effort going into the SAML standard. Working implementations are now being developed by the major Internet security companies.

3.9 XML Access Control

OASIS has recently created a new technical committee [15] to investigate and create an access control standard using XML. The expression of authorisation policies in XML against objects that are themselves identified in XML could be useful for advanced Web servers and operating systems. The aim is to achieve functionality able to represent current access control policies. One example is the Unix access control model, giving users and groups read, write and execute privileges to files. There is also the Java fine-grained access control to resources. It may be that access control depends to some extent on the protocols being used — a secure link is preferable to an open connection.

3.9.1 Multiple Layer Security

In determining a complete communication session between authenticating participants, a security structure that consists entirely of XML security mechanisms will be inefficient. Relying on XML Signature or XML Encryption for signing and encrypting all the messages passed could be a cumbersome and unwieldy process. XML should be used where it has an advantage over existing protocols. In many cases, end-to-end security using SSL (or IPsec) may be sufficient and more efficient. The new XML security mechanisms may find particular niches where they will be used. For example XML authentication tokens could be used at the beginning of a session, or XML Signature could be used at the end of a session for binding both parties to an agreement.

3.9.2 A New Security Architecture for the Web?

The new XML security mechanisms raise the possibility of a new Web architecture where security plays a key role (see Fig 3.7). Authentication and entitlement granting could be performed by Web services, enabling access control to other Web services. For example, a Web site dedicated to news and debate could have controlled permissions for publishing and accessing material. These may be granted by a number of authorisation Web services, one for journalists, one for those who wish to make comments, and a third for subscribers. Key revocation and registration issues can be dealt with using separate services that use the XKMS specification. There is certain to be components of existing PKI functionality that will be offered as Web services, but there is also the possibility that new services to support a

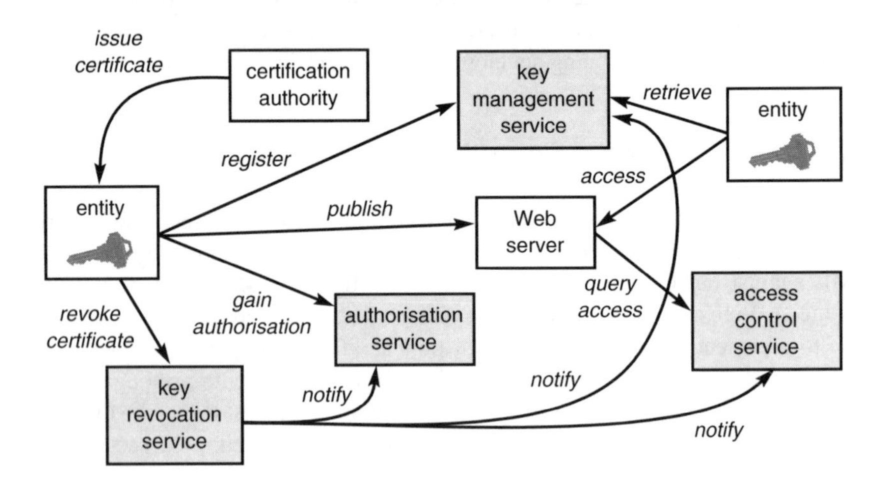

Fig 3.7 Interworking security Web services

public key infrastructure will emerge, such as attribute certificates and authorisation servers.

One key Web service for business-to-customer interaction would be a payments service. A customer, on making such a payment, could receive authorisation tokens enabling access to restricted services. It is likely that the communication between customer and payment service will be in XML, and the tokens will be an XML format, signed by the payments service.

In the end, only time will tell what ideas will prosper. Web services are too new and untested to be certain to succeed — so security mechanisms to support them also cannot be a safe bet. However, businesses that sell or require security should be aware of new possibilities in the future.

3.10 Summary

In the next few years, there will be a major effort at creating security services using XML. Eventually security at the XML level will settle down to a number of core areas, with other security necessities remaining at other layers, such as SSL or IPsec. However, the possibility of selling security Web services to secure business-to-business and business-to-customer eCommerce will be a major attraction to eSecurity and trust service providers.

References

1 SOAP version 1.1 (W3C note) — http://www.w3.org/TR/SOAP/

2 UDDI consortium — http://www.uddi.org/

3 WSDL (version 1.1) (W3C note) — http://www.w3.org/TR/wsdl

4 XAML — http://www.xaml.org/

5 ebXML (a joint initiative of UN/CEFACT (responsible for the EDI standard) and OASIS) — http://www.ebxml.org/

6 BT TrustWise — http://www.trustwise.com/

7 VeriSign — http://www.verisign.com

8 LDAP is standardised in RFC 2559 — http://www.ietf.org/rfc/

9 IETF: '*An Internet Attribute Certificate Profile for Authorization*', draft standard — http://www.ietf.org/internet-drafts/draft-ietf-pkix-ac509prof-06.txt

10 Kerberos (developed by the Massachusetts Institute of Technology) — http://web.mit.edu/kerberos/www/

11 OASIS — http://www.oasis-open.org/

12 XKMS (W3C note) — http://www.w3.org/TR/xkms/

13 Security services markup language — http://www.s2ml.org/

14 AuthXML — http://www.authxml.org/

15 XACML technical committee — http://www.oasis-open.org/committees/xacml/

4

SECURITY MODELLING LANGUAGE

M R C Sims

4.1 Introduction

This chapter introduces the Security Modelling Language (SecML) — an idea from BTexact Technologies. The SecML is a language, inspired by the Unified Modeling Language (UML), for specifying, visualising and documenting the security-related artefacts of a system. The language can be used to model systems at various levels of abstraction making it useful for tasks as diverse as designing key management protocols or analysing the threats against existing real-life systems. The SecML is mainly graphical and, although it could be written by hand or using general-purpose graphics tools, is intended to be written and drawn using software explicitly designed for the task. In this way any changes to elements in a model can automatically be reflected in the diagrams containing it. Context-sensitive menus can be used to edit and view additional information about elements not shown in the diagrams, such as software source code and design documentation.

This chapter does not present the full picture. Many of the ideas presented here are still in their infancy and some even conflict. Also, no formal definition of the language has yet been produced, but awaits a wider acceptance of the ideas. It is intended that this exposition will bring the ideas of the SecML to a wider audience, thus promoting discussion about its development and ideas about security modelling in general.

4.2 Why Model a System's Security?

Creating a good model for an information system is essential for its successful design, development, maintenance and management. This is especially true for complex large systems — the number of artefacts makes them difficult to understand and remember. As the complexity of a system increases, so does the importance of having a good model. A model is especially useful for communicating concepts to people who are unfamiliar with a system. Often a system's security is expressed using diagrams comprising notation of the architect's own making and generally, but informally, accepted notation. These diagrams are usually accompanied with a list of threats, risks and countermeasures in the form of a Security Policy Document (SPD). These documents can be, at best, tedious to read. They usually show relationships between the threats and countermeasures in the form of a table, but lack any real explanation of the associations between the mechanisms suggested. SPDs show what 'should' be implemented; they are usually accompanied by another document stating which mechanisms have actually been implemented and why some have been ignored. A model represented by formally agreed notation covering all aspects of a system's security would increase the accuracy of understanding without increasing complexity.

Modelling can also lead to reuse of design. If a model of a security solution is written with the appropriate level of abstraction, it can be reused in another situation. For example, the protocol describing RSA signing is independent of any of the implementing technology or the kind of information to be signed. This mechanism can be treated as a pattern and can be reused, even if it has to be implemented more than once.

4.3 Scope of the SecML

UML is a language that incorporates concepts of Booch, OMT and OOSE for the purpose of specifying, constructing, visualising and documenting a software-intensive system. It is an extremely flexible modelling language that simplifies the process of software design and management. UML has an extension mechanism through the use of stereotypes[1]. However, this provides only the ability to extend the meaning of existing UML diagrams and does not make them suitable for expressing security-related aspects. It is possible that some parts of the SecML could be reworked to become a formal extension to UML — this is outside the scope of this chapter. UML is cited here as it has been the main inspiration for the SecML — many of its aims for systems in general can be directly mapped on to the SecML's aims for the security-related aspects of systems. For this reason, many of the terms from UML have been adopted for the SecML.

The SecML has the following primary goals:

- to provide its users with a ready-to-use, expressive visual modelling language so that they can develop and exchange information about the security aspects of their systems;

[1] UML's stereotypes are a notation that can be used to extend the language. If UML's way-of-expressing a particular element is too restrictive when modelling a particular artefact, a stereotype can be attached to identify enhancement to the artefact's type.

- to provide a formal basis for analysing the security of a system;
- to be flexible enough for security to be expressed at various levels of abstraction;
- to provide a path linking the mechanisms employed by a system to the security requirements.

The author is particularly concerned that the SecML be convenient for everyday use as a means to help security engineers and system administrators design and manage real systems. Parts of the language should be extended or ignored depending on the formality of the context in which it is used.

4.4 Overview of the Modelling Approach

The SecML allows its users to represent systems as a collection of *elements* containing *attributes* bound to one another with *relationships*. Each element will have a *class* that defines the kinds of attribute and relationship it may have. Some elements are very easy to identify (such as computers and people), others are more difficult (such as the information on potentially malicious activity against the system). By drawing various kinds of diagram (different diagrams will be suitable for different domains), a user will be able to identify each of a system's elements. The SecML attempts to make it possible to express almost all of the relationships that have an impact on the security of the system without over-complicating the language. There will be times when the relationships do not allow the user to express exactly what or all that is required; when this happens, special documentation elements can be used.

For a more detailed explanation of the classes that comprise the meta-model, refer to 'Appendix A — Element Classes' and 'Appendix B — Element Class Diagrams'. The rest of this document assumes that the reader is familiar with these classes and their relationships. Whenever a relationship is referenced, the name will be shown in single quotes sometimes followed by the class of the target in parenthesis.

4.5 Building a Model

A user builds a SecML model mostly by drawing diagrams. These diagrams provide views on sub-sets of the elements of the whole model and help highlight particular relationships. The diagrams can be drawn by hand using pencil and paper or general-purpose drawing software. However, it is intended that special software be written to allow the diagrams to be drawn more easily and with the appropriate constraints imposed. Ideally the software will provide a strong linking between diagram components and the underlying elements. In this way a change made to an element or relationship in one diagram can cause automatic changes to other diagrams showing that element. For added convenience the software should also allow the user to navigate the relationships between elements using simple list-based windows. To keep the diagrams as simple as possible, complex graphics or the use of colour are not mandatory; however, tools for using the SecML could make use of these features.

The rest of this section looks at each of the diagram types. Textual descriptions and examples are used to describe the diagrams rather than formal definitions. As this is simply a presentation of the current state of the SecML rather than a formal specification, this format eases the understanding of the ideas and simplifies changes and the incorporation of new ideas. The diagrams use text to identify the artefacts being represented — this is mostly done by using the 'name' relationship of an Element which in many cases is constructed using the 'basic names' of other related elements.

Sometimes, due to the location or relative positioning of the text the full textual description is verbose. Because of this, whenever the 'name' of an Element is needed, a *context* will also be given. Tables C1—C4 in 'Appendix C — SecML Element Names' show how the 'name' is constructed for each of the element classes within different contexts, and 'Appendix D — Example Diagrams' provides three sample SecML diagrams.

For example, if we have a Realization, its full name (i.e. 'name' with empty context) might be:

{BankingSystem}AccountBalance|Bill:EndOfDay:encrypted[Server]

showing that the Realization is the *encrypted* copy of an Instance of *AccountBalance* (pertaining to *Bill*) called *EndOfDay* stored on the *Server*. If, from the context given in the diagram, we already know the InformationType, the System and its location, the 'name' will be given by:

!:EndOfDay:encrypted

It should be noted that Realizations will always have three colons — this is so they can be identified as Realizations. Exclamation marks ("!") are used in place of InformationTypes and Instances to draw a distinction between 'basic names' absent due to the context and empty 'basic names'. Often the 'basic name' of instances will be empty if the Instance represents a typical value. The 'basic name' of Realizations will only usually be non-empty if a particular state is to be described, or a distinction needs to be drawn between Realizations with the same 'owner' (Instance) and the same 'location'.

4.5.1 StaticDiagrams

StaticDiagrams allow a user to identify Components of a System and their Interfaces. At the top of the diagram (see Appendix D) is its 'name' (context: *empty*). The body of the diagram contains 'stick-men' representing Actors. Under each of these is the 'name' (context: 'system' of the diagram) of the Actor. Boxes are represented in a similar way but using a rectangle with the 'name' (context: 'system' of the diagram) written on the inside aligned with the top edge. Any of its 'sub-boxes' are drawn nested within the rectangle in a similar fashion. Beneath the 'name' of each Component is a String containing the first letters of the attribute values — 'R' or 'I' for existence, 'A' or 'G' for specificity, and 'P' or 'L' for corporeality. Note that no letter is used if an attribute is 'undefined'.

Interfaces are represented by lines connecting its 'components'. The 'name' (context: 'system' of the diagram and both 'components' of the Interface) of each Interface is written by the line near the mid-point.

Interfaces can also be shown when only one of its 'components' is shown in the diagram. In this case the other end of the line should connect to a side of the diagram. The 'name' (context: 'system' of the diagram and the 'component' of the Interface shown) is written at this end.

The positioning of the Actors and Boxes within the diagram can be used to help describe the physical or logical positioning of the components, but no rules are imposed.

4.5.2 ProtocolDiagrams

ProtocolDiagrams are used to describe a sequence of events and how the information is passed between Components during those events.

Each ProtocolDiagram is a view of a Scene (see Appendix D). At the top of the diagram is the 'name' (context: *empty*) of the Scene. Underneath this is the 'name' (context: 'system' of the Scene) of the diagram itself (which may be an empty String). Beneath this, but distributed across the diagram, are 'collaborators' (Components) of the Scene. These are represented in the same way as in StaticDiagrams (context: 'system' of the Scene). Boxes can be drawn so that the 'sub-boxes' are drawn in a horizontal row within the rectangle of the parent Box or outside the parent Box. A vertical line from each Component is drawn down the diagram. To the right of each Component are the 'names' (context: 'system' of the Scene and the Component itself) of any Realizations it 'contains' that have any bearing on the Scene. Below the Components and crossing the vertical lines are a number of areas (one below the other separated by dashed lines) to describe each of the 'activities' of the Scene. On the left side of each activity area (to the left of any vertical lines) is the 'name' (context: 'system' of the Scene and the Activity itself) of the Activity.

To the right of each vertical line within the activity area is the 'name' (context: 'system' of the Scene and the Component corresponding to the vertical line) of any 'inceptions' and 'terminations' of the Activity owned by the Component. Any Inceptions with 'derived from' equal to any of the Realizations shown will have a horizontal arrow drawn from that Realization's vertical line to the vertical line corresponding to the Inception. Similar arrows will be drawn from any 'owners' of 'transformations' that are equal to any of the Realizations shown.

4.5.3 InformationDependencyDiagrams

InformationDependencyDiagrams are used to describe how pieces of information depend on one another (see Appendix D). For example, an encrypted message will depend on the decryption algorithm and the key if it is ever to be read again. Note that an encrypted message would be represented by a particular Realization of that message's Instance, whereas any Realization of the algorithm or key (not encrypted or hidden in any way) will be suitable for the decryption. The diagrams also show threads of Inceptions indicating how Realizations become more concealed or revealed than the Realization used to derive them.

The primary graphics in the diagram are rectangles containing the 'names' of Realizations. The 'owner' (Instance) of Realizations can be represented as a rectangle with a double left edge connected using solid lines to its 'realizations'. This helps to reinforce the relationship between Realizations with the same 'owner' and reduces duplication when displaying names. The rectangles contain the 'names' of the instances. The 'owner' (InformationType) of Instances can be represented as a rectangle with a triple left edge connected using solid lines to its 'instances'. This helps to reinforce the relationships between Instances with the same 'owner' and, again, reduces duplication when displaying names.

The 'derived from' relationship of Inceptions are shown using solid arrows from the rectangle of the source Realizations to the 'owner' of the realization. These arrows can be labelled using the Inception's 'basic name'.

The contexts governing the 'names' of the InformationType, Instances and Realizations are described below.

- InformationType — the context includes:

  — the 'system' of the diagram.

- Instance — the context includes:

  — the 'system' of the diagram;

  — (if the Instance's rectangle is connected to the rectangle of its 'owner') the 'owner' of the Instance.

- Realization — the context includes:

  — the ‘system’ of the diagram;

  — (if the Instance’s rectangle is connected to the rectangle of its ‘owner’ or the Instance has a ‘derived from’ arrow coming from another Instance) the ‘owner’ of the Instance and the ‘owner’ of the ‘owner’ of the Instance;

  — (if the Instance’s rectangle is connected to the rectangle of the ‘owner’ of its ‘owner’) the ‘owner’ of the ‘owner’ of the Instance.

Inceptions that have ‘transformations’ can have these represented by a dashed line running from the Inception’s ‘owner’ rectangle to the rectangle of the ‘Realization’ of the Transformation itself. A small square will be put at the end point of the line (joining on to the rectangle) containing either “c” (for a ConcealingTransformation), “r” (for a RevealingTransformation) or “b” (for a BindingTransformation).

The ‘potential uses’ of Instances can be represented by dashed lines from the Instance to the ‘realization’ of the PotentiallyUsedTo. In a similar way to the above, a small square will be put at the end point of the line (joining on to the rectangle) containing either “pr” (for a PotentiallyUsedToReveal) or “pb” (for a PotentiallyUsedToVerifyBinding).

4.6 Summary

Readers of this chapter may be forgiven for thinking that modelling using a well-defined language is more effort than it is worth. However, the evidence from those designing and developing software shows that, although a modelling syntax can take a lot of time to become familiar with, it can accelerate development and reduce mistakes and misunderstandings. The adoption of a modelling language by those in the security community would almost certainly have a similar impact. The SecML is a language in its infancy that still needs to be changed to be more usable and to incorporate topics such as ‘threat analysis’ and ‘trust relationship’. Hopefully this chapter will help to encourage readers to think about how security models are built and how the security community can move closer to a globally accepted modelling language.

Appendix A

Element Classes

This appendix describes each of the classes of the meta-model. Classes that define relationships will show each in a separate table. The multiplicities are denoted using the usual UML notation; where *Composition* is used the multiplicity of the source is 1, but additionally the lifetime of the target is governed by the lifetime of the source. For some relationships the reverse reference (from target to source) may also be navigable — this is not shown. The relationship 'owner' is given to all reverse references where *Composition* is used. Primitive classes such as String are not defined.

Element (abstract)
Superclass: None
Description: This is the top-level abstract class.

Name	basic name
Source multiplicity	Composition
Target class	String
Target multiplicity	1
Description	Each element has a basic name that identifies it. It can comprise any sequence of ASCII characters in the range 0x20 to 0x7A excluding: "!", "+", "-", ",", "(", ")", ":", "[" and "]". See the definitions of the subclasses of element for a more detailed description of how the 'basic name' should be used and suggested conventions.

Name	basic name
Source multiplicity	Composition
Target class	String
Target multiplicity	1
Description	The 'name' of an Element will depend on the context in which it is being used. In many cases this will be the same as the 'basic name'. See 'Appendix C — SecML Element Names' for further details.

Documentable (abstract)
Superclass: Element
Description: This is a common superclass for many elements. Elements of subclasses own a number of Documents.

Name	documents
Source multiplicity	Composition
Target class	Document
Target multiplicity	*
Description	Each Documentable element owns a number of Documents that can be used to describe the element in more detail than is allowed by the relationships or can be expressed in the diagrams. It also provides a mechanism to attach arbitrary documents (such as SPDs, source code, management documents, etc) to an element.

Document (abstract)
Superclass: Element
Description: Documents are textual-based elements that can be used to describe elements. Documents do not appear explicitly within the SecML diagrams, but SecML tools should allow access to Documents via the GUI (e.g. using context-sensitive menus).

DocumentPointer (abstract)
Superclass: Document
Description: Documents that are stored outside a SecML model can be referred to using DocumentPointer.

URL (concrete)
Superclass: DocumentPointer
Description: Contains a URL String pointing to the actual document data. This is useful to build links between Documentables and existing documentation in the form of, say, Web pages.

DocumentInline (abstract)
Superclass: Document
Description: Documents that are stored as part of a SecML model are instances of subclasses of DocumentInline. As well as containing text and pictures, they will contain direct references (interleaved with the text) to other elements in the system or to elements in prerequisite systems via the 'references' relationship.

Name	references
Source multiplicity	*
Target class	SystemElement
Target multiplicity	*
Description	The target of this relationship can be any SystemElements with 'system' the same as the 'system' of the source or a 'prerequisite' of the 'system' of the source.

Hypertext (concrete)
Superclass: DocumentInline
Description: Contains RichText with references to other elements.

Model (concrete)
Superclass: Documentable
Description: Models are the container elements for all others. No elements refer to models via a composition relationship. Each model owns a number of Systems.

Name	systems
Source multiplicity	Composition
Target class	System
Target multiplicity	*
Description	

System (concrete)
Superclass: Documentable
Description: Each Model is composed of a number of Systems. Each System owns a number of SystemElements that describe the physical and logical components, the information and the processes. Each System can have a number of prerequisite systems whose owned elements can potentially be related to by its owned elements.

Name	direct prerequisites
Source multiplicity	*
Target class	System
Target multiplicity	*
Description	The targets must have the same 'owner' (Model) as the source. A target cannot have the source as one of its 'prerequisites'. That is, no circular relationships are allowed.

Name	prerequisites
Source multiplicity	*
Target class	System
Target multiplicity	*
Description	This relationship is derived from the 'direct prerequisites' relationship. The target comprises the 'direct prerequisites' of the source and all of their 'pre-requisites'.

Name	components
Source multiplicity	Composition
Target class	Component
Target multiplicity	*
Description	The physical and logical items that are either considered part of the system or have an impact on the system (including people, computers, safes, etc).

Name	informationTypes
Source multiplicity	Composition
Target class	InformationType
Target multiplicity	*
Description	

Name	interfaces
Source multiplicity	Composition
Target class	Interface
Target multiplicity	*
Description	These denote boundaries between Components over which information can flow, but do not in themselves constitute Components.

Name	scenes
Source multiplicity	Composition
Target class	Scene
Target multiplicity	*
Description	These show the protocols within the system. The target of this relationship can contain both ScriptedScenes and UnscriptedScenes.

Name	diagrams
Source multiplicity	Composition
Target class	Diagram
Target multiplicity	*
Description	

SystemElement (abstract)
Superclass: Documentable
Description: All subclasses will have a derived relationship 'system' identifying the System owning the Element.

Name	system
Source multiplicity	1
Target class	System
Target multiplicity	1
Description	This relationship is derived. If the 'owner' of the source is a System, then it is also the target, else the target is the 'system' of the source's 'owner'

Component (abstract)
Superclass: SystemElement
Description: Components are physical or logical items in a system. They have the following attributes.
Existence — which can be one of (Undefined | Real | Imaginary). This is used to indicate how real the component is. A system administrator might be Real as it is a defined part of the system. An attacker, however, might be Imaginary as it may or may not exist.
Specificity — which can be one of (Undefined | Actual | General). This is used to show if the component represents an actual item in the system or a typical item. A server might be Actual as we may specify the make and model. A client, however, might be General and represent a typical client.
Corporeality — which can be one of (Undefined | Physical | Logical). This is used to draw a distinction between components that represent solid matter such as a computer and components that do not, such as an e-mail inbox.

Actor (concrete)
Superclass: Component
Description: Actors are simply people. However, if components are abstract they might represent both people and computers, say. In these cases Boxes should be used.

Box (concrete)
Superclass: Component
Description: Boxes are components that are not Actors. They have the additional property that they can have a number of sub-boxes.

Name	sub-boxes
Source multiplicity	1
Target class	Box
Target multiplicity	*
Description	The targets of this relationship can be any Boxes with 'system' the same as the 'system' of the source or a 'prerequisite' of the 'system' of the source.

Interface (concrete)
Superclass: SystemElement
Description: Interfaces identify boundaries between pairs of components over which information can move, but do not in themselves constitute components. Examples include, User↔PC (using a screen and keyboard), PC↔LAN (using the network card and software) and PaperDocument↔User (using pen and eyes).

Name	components
Source multiplicity	*
Target class	Component
Target multiplicity	2
Description	The targets of this relationship can be any Components with 'system' the same as the 'system' of the source or a 'prerequisite' of the 'system' of the source.

InformationType (abstract)
Superclass: SystemElement
Description: An InformationType shows its existence as a piece of information in the system. It does not identify actual values but is a template for instances. The convention is to start the 'basic name' of each InformationType with a capital letter, use no white-space and be non-empty.

Name	instances
Source multiplicity	Composition
Target class	Instance
Target multiplicity	*
Description	Targets represent particular values of the source. For example, an InformationType with 'basic name' DataEncryptingKey might have the Instances Old and New when considering a key updating process, another Instance with empty 'basic name' when considering the encryption process.

Name	pertains to
Source multiplicity	*
Target class	Component
Target multiplicity	*
Description	Each InformationType can have a number of components about which it relates. The targets of this relationship can be any components with ‘system’ the same as the ‘system’ of the source.

Name	pertains to string
Source multiplicity	*
Target class	String
Target multiplicity	*
Description	This relationship is derived from the ‘pertains to relationship’ and purely exists for the calculation of the ‘name’ of the source and Elements owned by the source. If the target of ‘pertains to’ is empty then the target of this relationship is “ ”. Otherwise, the target is defined as: “|” <A “,” separated list of ‘names’ of the targets of ‘pertains to’>

AtomicInformationType (concrete)
Superclass: InformationType
Description: This is a simple InformationType whose component parts will not be considered separately. In many systems, cryptographic keys will be atomic; in some, however, the component bits, bytes or words may need to be modelled separately due to the nature of the possible attacks on the system.

CompositeInformationType (concrete)
Superclass: InformationType
Description: CompositeInformationTypes provide a means to express structures or compositions of other InformationTypes.

Name	composed of
Source multiplicity	*
Target class	InformationType
Target multiplicity	2..*
Description	The targets and the source must have the same ‘system’.

Instance (concrete)
Superclass: SystemElement
Description: An Instance represents an actual occurrence or edition of a piece of information. Usually no real values are considered but stereotypes. The convention for the 'basic name' is to start with a capital letter and have no white-space. When the Instance represents any stereotypical value, the 'basic name' can be empty. Other times it is convenient to draw a distinction between Instance of the same InformationType. For example, when considering the changing of a password (InformationValue 'basic name' "Password") it might be suitable to have two Instances with 'basic names' "OldPassword" and "NewPassword".

Name	worth
Source multiplicity	Composition
Target class	Worth
Target multiplicity	1

Name	realizations
Source multiplicity	Composition
Target class	Realization
Target multiplicity	*
Description	The targets represent copies or occurrences of the source. An Instance with 'basic name' "NewPassword" might have three Realizations, one in the memory of an Actor, one in plain-text in the memory of a PC and one encrypted on the hard-disk of the same PC.

Name	potential uses
Source multiplicity	Composition
Target class	PotentiallyUsedTo
Target multiplicity	*
Description	

Worth (concrete)
Superclass: SystemElement
Description: Worth is an indication of value. How this will be represented in the SecML is yet to be determined. It will probably give some indication as to the importance of confidentiality, integrity and availability of the 'owner' (Instance).

Realization (concrete)
Superclass: SystemElement
Description: A Realization is a copy or occurrence of an Instance. A Realization is not necessarily in a form that makes the raw value available. For example, a customerAccount Realization could be encrypted using an algorithm and key. This encrypted version would be a Realization of the same Instance as the original. Similarly, a combination for safe or steganographic techniques could be used to conceal the information.

Often Realizations will have an empty 'basic name' as the naming of its 'owner' (Instance) and its 'location' are enough to identify it. However, the 'basic name' can be used as an indication as to the state of the Realization (e.g. "encrypted").

Name	location
Source multiplicity	*
Target class	Component
Target multiplicity	1
Description	Each Realization must be located in a component. If information moves between components, say from a Smartcard to a PC, this would be represented using multiple Realizations. The Activity class would be used to show how the Instances are derived from one another. The target of this relationship can be any Components with 'system' the same as the 'system' of the source or a 'prerequisite' of the 'system' of the source.

Name	inception
Source multiplicity	Composition
Target class	Inception
Target multiplicity	0..1
Description	

Name	termination
Source multiplicity	Composition
Target class	Termination
Target multiplicity	0..1
Description	

PotentiallyUsedTo (abstract)
Superclass: SystemElement
Description: Each Instance can potentially be used to reveal verify the binding of a number of Realizations. For example, a Realization 'EncryptedData' that was encrypted (concealed) using a particular Realization of the Instance 'SessionKey' could potentially be decrypted (revealed) with any other instance non-concealed Realization of 'SessionKey'.

Name	realization
Source multiplicity	*
Target class	Realizations
Target multiplicity	1
Description	The target of this relationship can be any Realization with 'system' the same as the 'system' of the source or a 'prerequisite' of the 'system' of the source.

PotentiallyUsedToReveal (concrete)
Superclass: PotentiallyUsedTo
Description: The owning Instance can potentially be used to reveal the target of the 'Realization' relationship. The term 'reveal' is used to mean decrypt, but it also covers non-cryptographic ways of recovering information that has been hidden.

PotentiallyUsedToVerifyBinding (concrete)
Superclass: PotentiallyUsedTo
Description: The 'owner' (Instance) can potentially be used to verify the binding of the target of the 'realization' relationship. The term "verify the binding" is used to mean check a digital signature or cryptographic hash. It also covers non-cryptographic ways of showing pieces of information have been bound together.

Termination (concrete)
Superclass: SystemElement
Description: A Termination shows the end of the life of a Realization. Each Realization can have zero or one Terminations. Terminations can also be referenced as an Activity showing the events that lead to the end of the Realization.

Inception (concrete)
Superclass: SystemElement
Description: An Inception represents the creation of its 'owner' (Realization). Creating a Realization can happen in one of three ways:

- be a simple copy of another Realization with the same 'owner' (Instance) — this might occur as information moves from one Component to another;
- be a transformed version of another Realization with the same owner — the transformation might take the form of encryption;
- be the first Realization of the 'owner' (Instance).

Name	derived from
Source multiplicity	*
Target class	Realizations
Target multiplicity	0..1
Description	If the creation of the Realization is of the third type above, then there will be no target for this relationship. Otherwise it is a Realization with 'system' the same as the 'system' of the source of a 'prerequisite' of the 'system' of the source. The 'owners' of the target and source must also be the same.

Transformation (abstract)
Superclass: SystemElement
Description: Subclasses of this class show how another Realization has been used to transform a Realization during its Inception.

Name	realization
Source multiplicity	*
Target class	Realization
Target multiplicity	1
Description	The target is the Realization used to transform the 'owner' of its 'owner' (Realization) during its inception.

BindingTransformation (concrete)
Superclass: Transformation
Description: The 'realization' has been used to bind the 'owner' of the 'owner' during its inception. The term "bind" is used as a generic term for preserving integrity. This could cover techniques such as the attachment of a keyed hash, or physical techniques such as written signatures or wax seals.

ConcealingTransformation (concrete)
Superclass: Transformation
Description: The 'Realization' has been used to conceal the 'owner' of the 'owner' during its inception. The term "conceal" is used as a generic term for hiding or disguising the meaning. This could cover techniques such as encryption or steganography.

RevealingTransformation (concrete)
Superclass: Transformation
Description: The 'Realization' has been used to reveal the 'owner' of the 'owner' during its inception. This is the inverse process to concealing. However, if information has undergone many ConcealingTransformations (e.g. multiple encryption) a single RevealingTransformation will not necessarily bring the information back to a state where it has meaning, but just undo one of the concealing layers.

Scene (abstract)
Superclass: SystemElement
Description: Scenes contain a sequence of Activities showing a process or protocol in a System.

Name	activities
Source multiplicity	Composition
Target class	Activity
Target multiplicity	* {ordered}
Description	

Name	collaborators
Source multiplicity	*
Target class	Component
Target multiplicity	*
Description	The target of this relationship is derived. It is the 'locations' of the 'owner' (Realization) of both the 'inceptions' and 'terminations' of the 'activities' of the source.

ScriptedScene (concrete)
Superclass: Scene
Description: A scripted scene is a process or protocol that is carried out under the authorisation of the system administrator. These scenes might be an integral part of the system or part of a solution to protect the system.

UnscriptedScene (concrete)
Superclass: Scene
Description: UnscriptedScenes represent attacks or other failures occurring in the system. They help to show the consequences of such an event.

Activity (concrete)
Superclass: SystemElement
Description: Activities are the steps within a protocol. For each step an optional 'executor' is identified.

Name	inceptions
Source multiplicity	0..1
Target class	Inception
Target multiplicity	*
Description	The Inceptions that occur during the Activity. The target of this relationship can be any Inceptions with 'system' the same as the 'system' of the source or a 'prerequisite' of the 'system' of the source.

Name	terminations
Source multiplicity	0..1
Target class	Termination
Target multiplicity	*
Description	The Terminations that occur during the Activity. The target of this relationship can be any Terminations with 'system' the same as the 'system' of the source or a 'prerequisite' of the 'system' of the source.

Name	executor
Source multiplicity	*
Target class	Component
Target multiplicity	0..1
Description	The target is the Component responsible for the execution of the source. The target of this relationship can be any Component with 'system' the same as the 'system' of the source or a 'prerequisite' of the 'system' of the source.

Diagram (abstract)
Superclass: SystemElement
Description: Diagrams help express other relationships between elements in a System (or prerequisite System).

StateDiagram (concrete)
Superclass: Diagram
Description: Identifies the Interfaces between Components.

Name	interfaces
Source multiplicity	*
Target class	Interfaces
Target multiplicity	*
Description	The target must be a subset of the 'interfaces' of the 'system' of the source and the target of the 'interfaces' of the 'prerequisites' of the 'system' of the source.

Name	components
Source multiplicity	*
Target class	Component
Target multiplicity	*
Description	The target must be a subset of the 'components' of the 'system' of the source and the target of the 'components' of the 'prerequisites' of the 'system' of the source.

ProtocolDiagram (concrete)
Superclass: Diagram
Description: Provides a view of a Scene.

Name	scene
Source multiplicity	*
Target class	Scene
Target multiplicity	1
Description	The target of this relationship can be any Scene with 'system' the same as the 'system' of the source or a 'prerequisite' of the 'system' of the source.

Name	activities
Source multiplicity	*
Target class	Activity
Target multiplicity	*
Description	The target must be a subset of the 'activities' of the 'scene' of the source.

Name	inceptions
Source multiplicity	*
Target class	Inception
Target multiplicity	*
Description	The target must be a subset of the 'inceptions' of the 'activities' of the source.

Name	terminations
Source multiplicity	*
Target class	Termination
Target multiplicity	*
Description	The target must be a subset of the 'terminations' of the 'activities' of the source.

InformationDependencyDiagram (concrete)
Superclass: Diagram
Description: Shows how InformationTypes, Instances and Realizations depend on each other.

Name	informationTypes
Source multiplicity	*
Target class	InformationType
Target multiplicity	*
Description	The target must be a subset of the 'informationTypes' of the 'system' of the source and the 'informationTypes' of the 'prerequisites' of the 'system' of the source

Name	instances
Source multiplicity	*
Target class	Instance
Target multiplicity	*
Description	The target must be a subset of the 'instances' of the 'informationTypes' of the 'system' of the source and the 'instances' of the 'informationTypes' of the 'prerequisites' of the 'system' of the source.

Name	realizations
Source multiplicity	*
Target class	Inception
Target multiplicity	*
Description	The target must be a subset of the 'realizations' of the 'instances' of the 'informationTypes' of the 'system' of the source and the 'realizations' of the 'instances' of the 'informationTypes' of the 'prerequisites' of the 'system' of the source.

Name	transformations
Source multiplicity	*
Target class	Transformation
Target multiplicity	*
Description	The target must be a subset of the 'transformations' of the 'inceptions' of the 'realizations' of the 'instances, of the 'informationTypes' of the 'system' of the source and the 'transformations' of the 'inceptions' of the 'realizations' of the 'instances' of the 'informationTypes' of the 'prerequisites' of the 'system' of the source.

Name	potential users
Source multiplicity	*
Target class	PotentiallyUsedTo
Target multiplicity	*
Description	The target must be a subset of the 'potential uses' of the 'instances' of the 'informationTypes' of the 'system' of the source and the 'potential uses' of the 'realizations' of the 'instances' of the 'informationTypes' of the 'prerequisites' of the 'system' of the source.

Appendix B

Element Class Diagrams

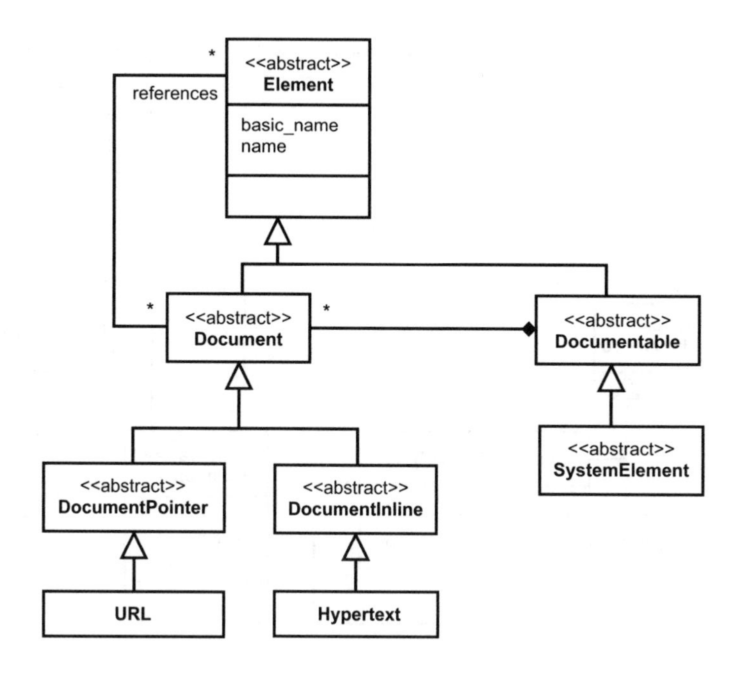

Fig 4.B1 Documentation Relation Classes.

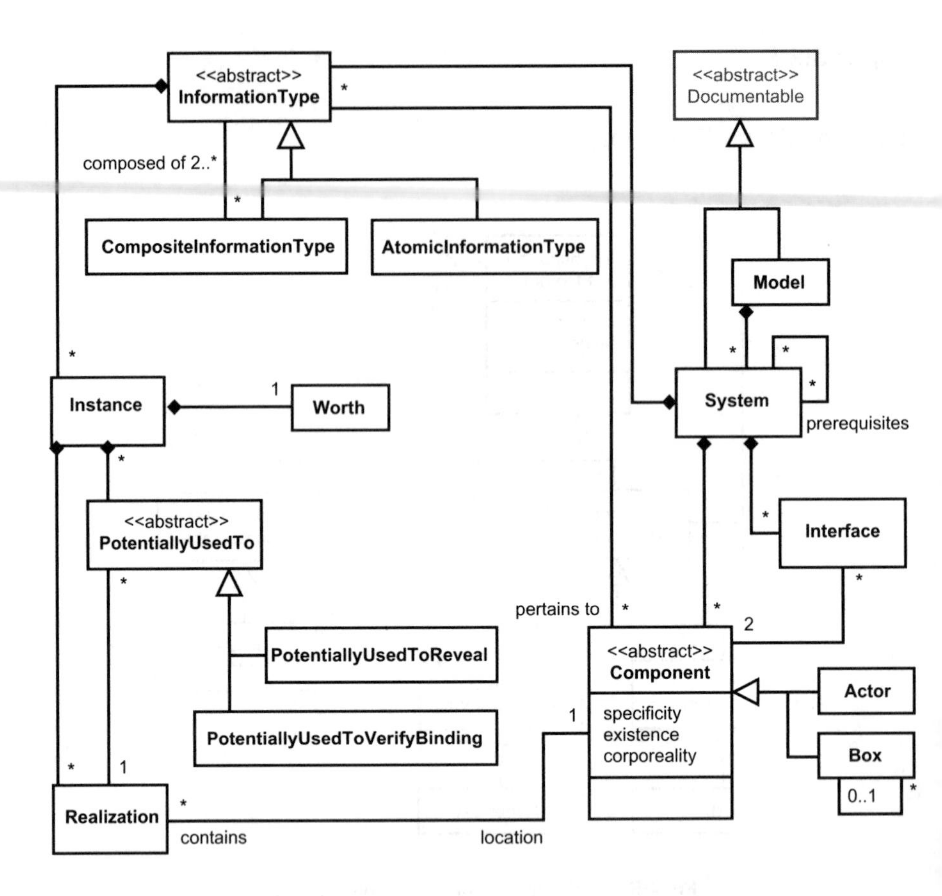

Fig 4.B2 Information and Component Related Classes.

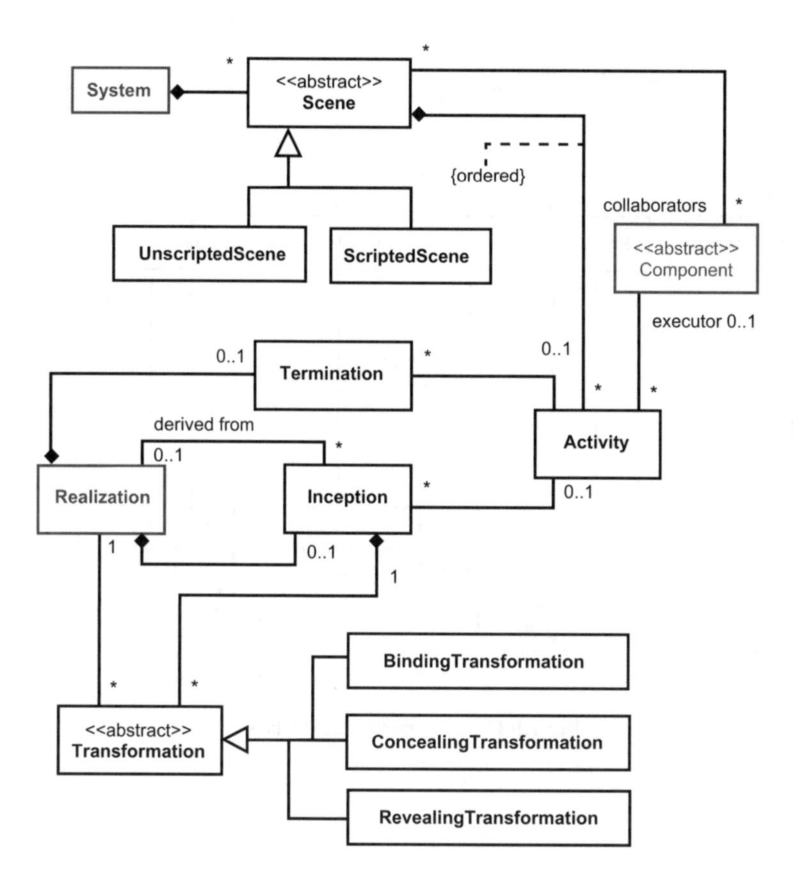

Fig 4.B3 Activity Related Classes.

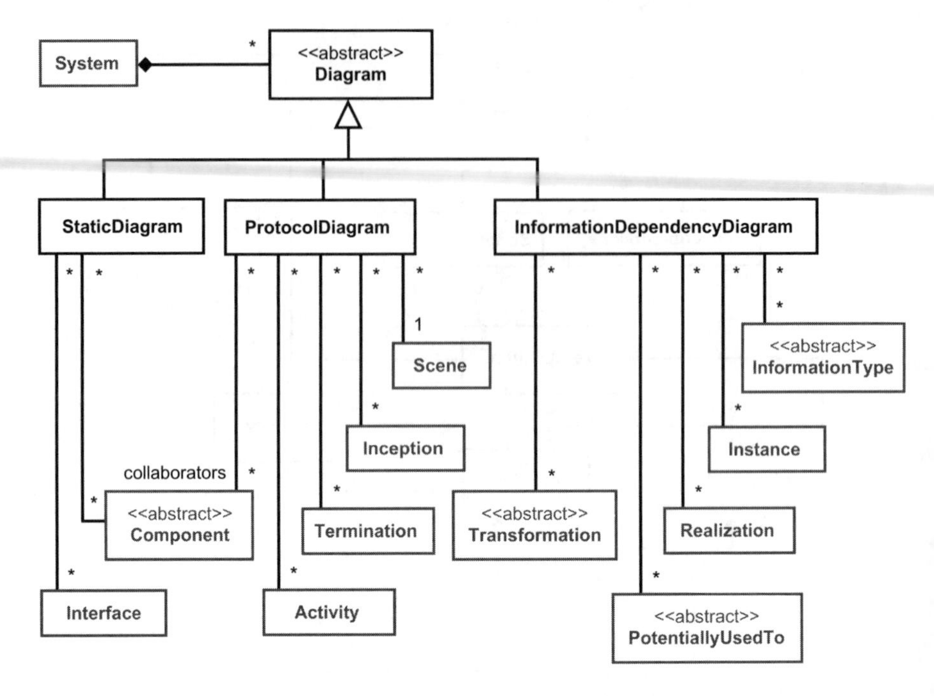

Fig 4.B4 Diagram Related Classes.

Appendix C

SecML Element Names

This appendix defines how Elements can be expressed as a String. Each Element has a ‘basic name’ but this is often not enough to describe the element without being verbose. Each Element also has a ‘name’ attribute but, unlike the ‘basic name’, this is dependent on the context in which the name is used. Whenever this document refers to the ‘name’ of an Element the context will be specified. The ‘name’ of an element comprises four parts:

<Prefix 1> <Prefix 2> <Basic Name> <Postfix>

Tables C1—C4 give the definition of each of these parts for each of the Element Classes. If a Class is not included in a table then the definition of the superclass should be used.

Table 4.C1 Prefix 1.

Class Name	**Condition**	**String Part**
Element	True	None
SystemElement	The context includes the ‘system’ of the SystemElement	None
	The context does not include the ‘system’ of the SystemElement.	“{” <The ‘basic name’ of the ‘system’> “}”
Inception	True	“+”
Termination	True	“–”

Table 4.C2 Prefix 2.

Class Name	Condition	String Part
Element	True	None
Instance	The context includes the 'owner' (InformationType) of the Instance.	"!:"
	The context does not include the 'owner' (InformationType) of the Instance.	<The 'basic name' of the 'owner'> <The 'pertains to string' of the 'owner'> ":"
Realization	The context includes the 'owner' (Instance) and the 'owner' of the 'owner' (InformationType) of the Realization	"!:!:"
	The context does not include the 'owner' (Instance) but does include the 'owner' of the 'owner' (InformationType) of the Realization.	"!:" <The 'basic name' of the 'owner'> ":"
	The context does not include the 'owner' (Instance) or the 'owner' of the 'owner' (InformationType) of the Realization.	<The 'basic name' of the 'owner' of the 'owner'> <The 'pertains to string' of the 'owner'> ":" <The 'basic name' of the 'owner'> ":"

Table 4.C3 Basic name.

Class Name	Condition	String Part
Element	True	The 'basic name' of the Element.

Table 4.C4 Postfix.

Class Name	**Condition**	**String Part**
Interface	The context includes neither or only one of the ‘components’ of the Interface	“(” <“,” separated list of the ‘basic names’ of the ‘components’ not included in the context> “)”
Realization	The context does not include the ‘location’ (Component) of the Realizations	“[” <The ‘basic name’ of the ‘location’> “]”
Inception	The Inception has a ‘derived from’ with multiplicity of zero	“(” <A “,” separated list of the ‘names’ of the ‘realizations’ of the ‘transformations’ > “)”
	The Inception has a ‘derived from’ with multiplicity of one	“(” <The ‘name’ of the ‘derived from’ Realization> “|” <A “,” separated list of the ‘names’ of the ‘realizations’ of the ‘transformations’ with whitespace>“)”

Appendix D

Example SecML Diagrams

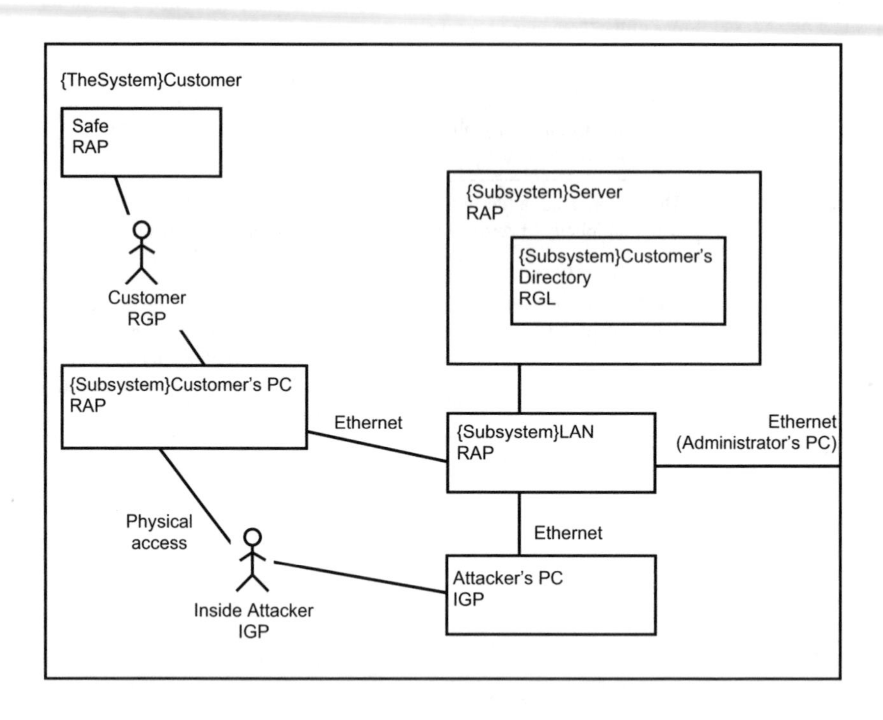

Fig 4.D1 Example Static Diagram.

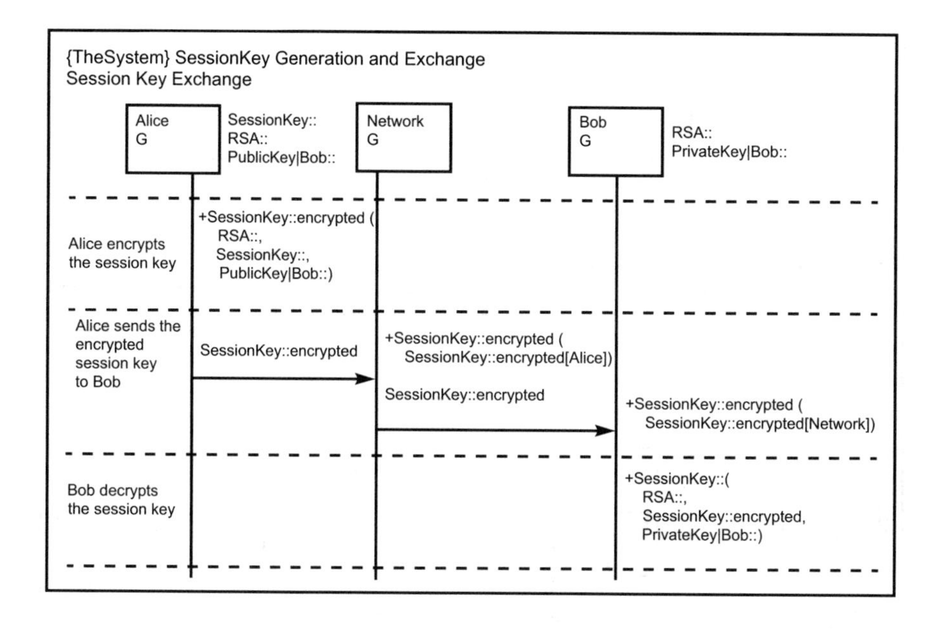

Fig 4.D2 Example Protocol Diagram.

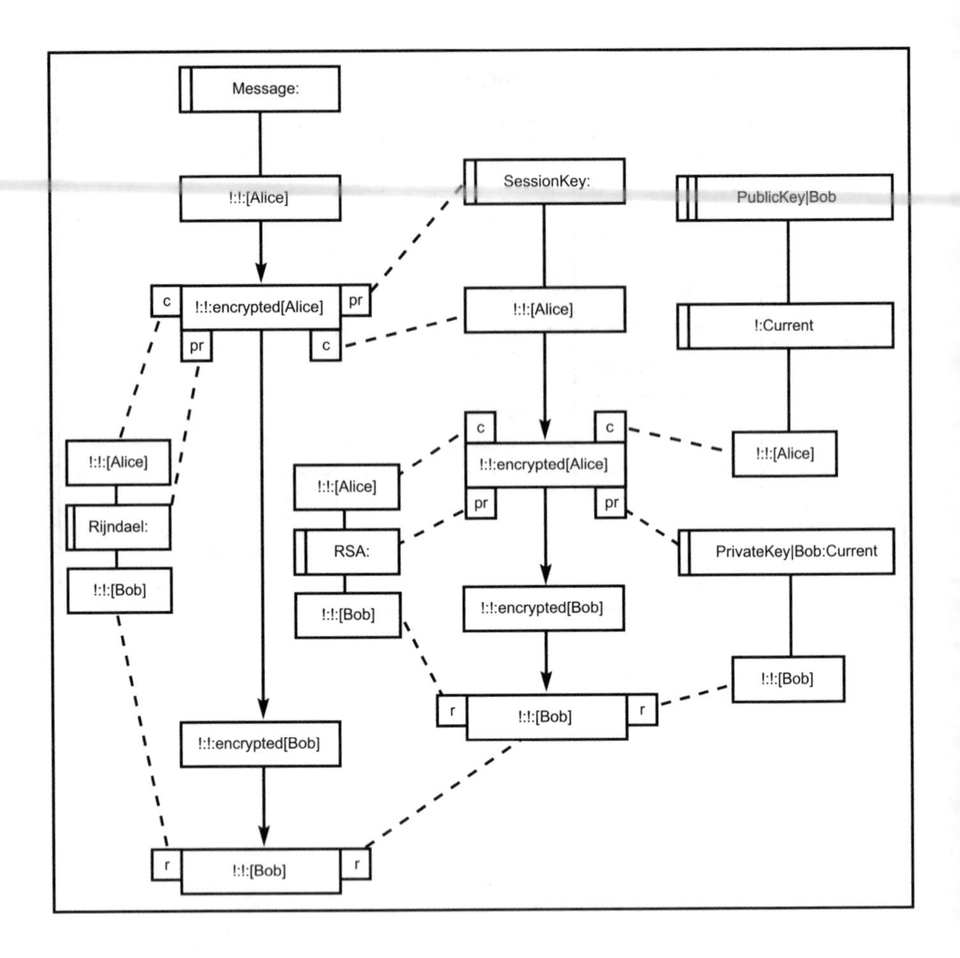

Fig 4.D3 Example Information Dependency Diagram.

5

PUBLIC KEY INFRASTRUCTURES — THE NEXT GENERATION

K P Bosworth and N Tedeschi

5.1 Introduction

In answer to the question: 'So when are PKIs actually going to happen, then?', the simple reply is that they already have. They are well established and, indeed, not only is there a thriving PKI industry occupied by both technology suppliers and service suppliers, but these established high-technology companies are quickly consolidating to gain market share.

A better question to ask might be: 'Is a contemporary PKI a technology or a service?' The answer is the latter. PKI suppliers are positioning themselves to offer trusted certification authority services that allow their customers to procure bolt-on PKI security components. This permits application and service providers to focus on their strengths in their own markets.

Although the whole field of public key infrastructure (PKI) still has many unanswered, and very complex, problems to be resolved, the most fundamental misconception is the basic definition of a PKI — it is a public key 'infrastructure'. This highlights the fact that the cryptographic technology is relatively unimportant and that a PKI as an infrastructure is not, and never will be, a stand-alone application in its own right.

The main strength of a PKI is that it can offer a scalable security solution component for any of the various connectivity layers (application, transport and network). A PKI is powered by cryptography that is under constant critical public review regarding algorithms and key sizes while also enjoying significant visibility within the global legal communities. Many national and international digital signature laws, and directives, such as the Electronic Communications Act (ECA), ESIGN, the Uniform Electronic Transaction Act (UETA) and the European Digital Signature Directive are now being ratified and enforced. It would be easy to believe that, if your digital assets have value, then they are certifiable.

So, like the first automobiles (more aptly described as horseless carriages in their time), the PKI today represents an engine (of sorts) bolted on to a cart in place of a horse. The overall vehicle works but it is usually expensive, unreliable, misunderstood (or even feared), difficult to install or maintain, and not very user friendly to operate.

Despite good basic principles and standards for guidance, current PKI solutions are highly proprietary. A PKI does, however, work quite well in this far-from-ideal environment but a lot of evolution and problem-solving needs to take place to arrive at the equivalent of the modern motor car with which we are familiar in the 21st century. This is a truly sophisticated, versatile vehicle that can meet the strict and varied demands for safety and emissions around the globe but can be easily driven by almost anybody on any highway. It is a vehicle that can be easily manufactured in both left and right hand drive, manual and automatic transmission, petrol, diesel and electric forms of power. It is certain that PKIs will advance to a similar degree — and hopefully at Internet evolutionary speeds.

This chapter does not attempt to explain the cryptography or the digital signature methods used in conjunction with a PKI. This is to say, it does not cover the current, or future, applied uses of the actual cryptographic keys that are assured via the use of a PKI (for example, signing e-mails, authenticating users to secure Web sites, controlling VPN tunnels). Nor does it go into great detail about the components within the PKI itself.

It does, however, concentrate on the issues surrounding the operation, usability and scalability of PKIs in a very real commercial world both for today and into the future. Looking forward, a number of interesting developments are appearing on the horizon. Validation authorities (VAs), validation brokers (VBs) and greater protocol support at the application, transport and network connectivity layers are good examples.

5.2 A Brief History of Public Key Infrastructures

This section describes the history of PKI evolution up to the present day. It brings the reader up to date with all the necessary background about PKIs and their current implementations and use.

5.2.1 Why it all began

Almost as soon as the mathematical possibilities of public/private key cryptography were recognised, and understood, a problem began to emerge. The elegance of the two-key asymmetric cryptography mechanism (whereby one key is used to encrypt, and the other is used to decrypt) quickly got stuck on a practical implementation issue. Although a key-pair owner could jealously guard their private key and freely publish their public key, it was impossible to prove that the published key really belonged to the person who claimed it as their own. What was needed was a trusted method of permanently 'binding' the owner's identity to the public key. So the trust hierarchy was conceived whereby a point of trust would bind a public key to an identity (and possibly include some other information too) on behalf of the owner of the key-pair. Everybody could then choose to accept the single point of trust as a reliable binding authority.

The point of trust needs to act responsibly and validate each application carefully before it performs the binding operation (otherwise people would quickly lose all faith in the correctness of the bindings that the trust-point performed). Today, validation usually focuses on whether you are who you say you are (as opposed, for example, to whether you are good for a certain amount of credit, or empowered in a particular financial authority role). Validation is performed using out-of-band (OOB) enquiry processes either directly with the applicant or with a third party such as Experian, Dunn and Bradstreet or the UK Companies House. The degree of validation can be reflected using additional information contained in the binding. This gives rise to a tiered 'class' structure for certificates (see Table 5.1).

Table 5.1 Levels of validation checking performed before a certificate is issued.

Class	Applicant, or their representing agent, is contacted	Third party is contacted	Applicant attends in person with formal documentary proof	Notes (All classes require that the enrolment information, especially the distinguished name, is unique)
1				Checks that a valid e-mail address is part of the enrolment.
2	YES			Usually for PRIVATE hierarchies. May use an HR record.
3	YES	YES		Usually a company validation. Includes a check that the application is empowered to make the enrolment on the company's behalf.
4		YES	YES	Face-to-face attendance with supporting formal documents. Notarisation of the meeting and further checking of the paperwork with the issuing agency (e.g. passport).

The cost of the binding operation can be proportional to the amount of validation effort involved and thus a workable commercial model began to emerge. Companies set themselves up to offer this service. Some companies even considered how the owners might want to use their shiny new cyberspace 'trusted identities' and began working with software companies to integrate the supporting functionality that was required.

In the early days, as with any pioneering 'gold-rush' situation, it was inevitable that a lot of hard work, time and money was spent trying to find what was actually only a very small amount of gold. The only people who stood to make any serious money were those who sold the pick-axes, shovels and whisky to the prospectors.

5.2.2 Certification Authorities

The correct terminology can now be introduced for use in the rest of this chapter. The trust points, mentioned above, are actually called certification authorities (CAs). Bindings of public keys to identities are called 'digital certificates' (or just 'certificates' herein). Key-pair owners are called 'end entities' (EEs). Sending a key and identity to a CA for certification is called 'enrolment'. Following validation of enrolment information, the result can be either an 'approval' or a 'rejection'. Mechanically creating a certificate from approved enrolment details is called 'signing'. Releasing a signed certificate to its rightful EE is called 'issuing'.

To allow a CA to prove its rightful ability to use its own private key to 'sign' a certificate for an EE, it creates its own 'root CA certificate' (Fig 5.1). This contains the CA's public key, identity, etc, but is usually 'self-signed' with its own CA private key. This means that the 'chain' of trust can stop at this point.

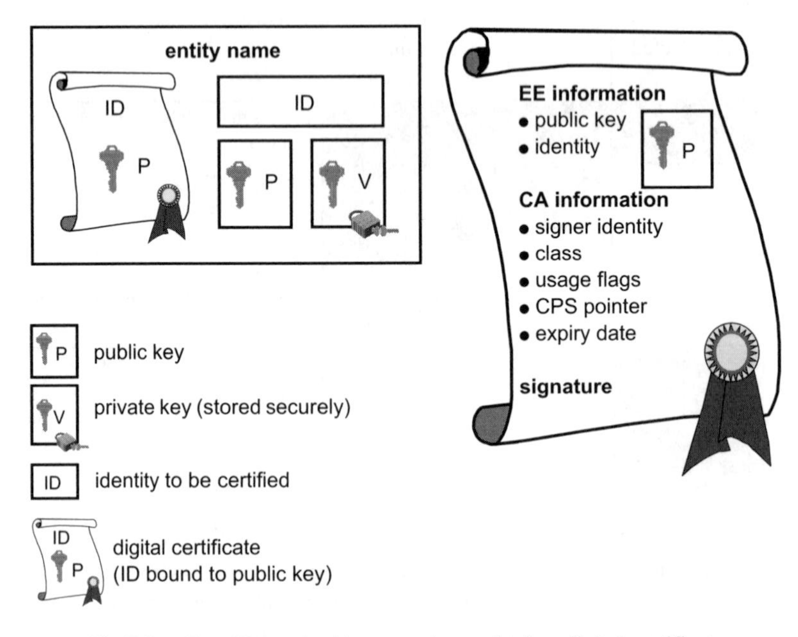

Fig 5.1 Details required by an entity to obtain a digital certificate.

5.2.3 Public and Private Hierarchies

A PKI can be built for two main operational scenarios. Firstly, it may be a public PKI whereby the CA root is widely recognised in the public domain (extensively published, or even included directly as part of the configuration of an application). Secondly, it might be a private PKI with limited visibility within a specific community of users.

The PKI is essentially a pyramid of trust (POT) structure with the CA root at the top and its signed EE certificates (sometimes called 'leaf' certificates) at the bottom. It is also possible to have intermediate layers within the POT whereby a CA can partition its certificates (perhaps by different types of intended usage) under a single CA root. Some CAs go one stage further and operate multiple CA roots to separate the POTs into their own different individual hierarchies (perhaps by class of certificate). Thus a CA might appear as a small number of public hierarchies plus a larger (and ever growing) number of private hierarchies for use only by individual customer-controlled communities.

In the interest of simplicity, this document treats a hierarchy as a single self-signed CA root certificate that directly signs its EE certificates. As an example, Fig 5.2 shows two independent POTs in which two self-signed CA root certificates have been used to sign their own separate groups of EE certificates. Remember that the POTs might be operated by different CAs, or they might represent different class hierarchies operated by a single CA.

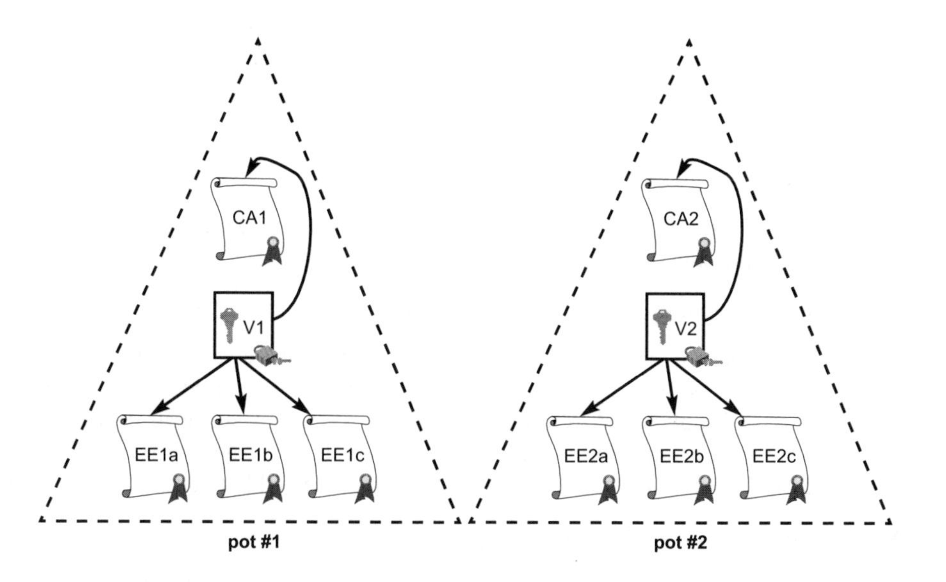

Fig 5.2 Two independent examples of the simple pyramid of trust that visually represents a PKI hierarchy.

A good example of the necessity to use both types, simultaneously, would be for a corporate Web server. The server would have a public hierarchy server certificate so that a standard browser application could recognise it and, therefore, trust it (this makes the job of building, and maintaining, corporate PCs easier). Corporate users who are permitted to log on to the server (to retrieve private corporate information) would then be issued with EE certificates under a private hierarchy (not necessarily even from the same CA). The private CA root only needs to be recognised by the server in order to limit its trust to employees who possess EE certificates within the same private hierarchy.

This process assumes that there is a suitable protocol for verifying the credentials of both parties at both ends of the connection and could be achieved by using a secure sockets layer (SSL) type of connectivity.

Note — do not confuse the use of the words 'public' and 'private' when applied to CA hierarchies, with their use when referring to public/private key pairs in the context of cryptographic technology.

5.2.4 Other PKI Components

There are many other parts required, within the CA architecture (Fig 5.3), to fully implement a PKI, which have yet to be covered. The main ones are described below.

- The registration authority (RA)

  This is the entity that performs the validation checking and makes the decisions whether to approve or reject enrolments. The RA can be an automatic process or a manual interface to a work flow system. The ability to perform RA functions is usually considered an important role within the CA and is securely controlled.

- The certificate database (CDb)

  This securely stores information about all the certificates that have been issued by the CA and supports rules to prevent duplicate identities being created.

- The certificate policy (CP) and certificate practices statement (CPS)

  The CP defines the high-level policy statements under which a CA will operate. The CPS defines the processes and procedures that will be used by the CA to deliver the CP policies. A reference to the CPS is usually included in the EE-certificate supporting information. Some of the complex issues surrounding the legal significance of CP and CPS documents are addressed later on.

- The enrolment/delivery interface

  This provides an EE with a means to submit enrolment information for certification and receive a certificate if it is approved.

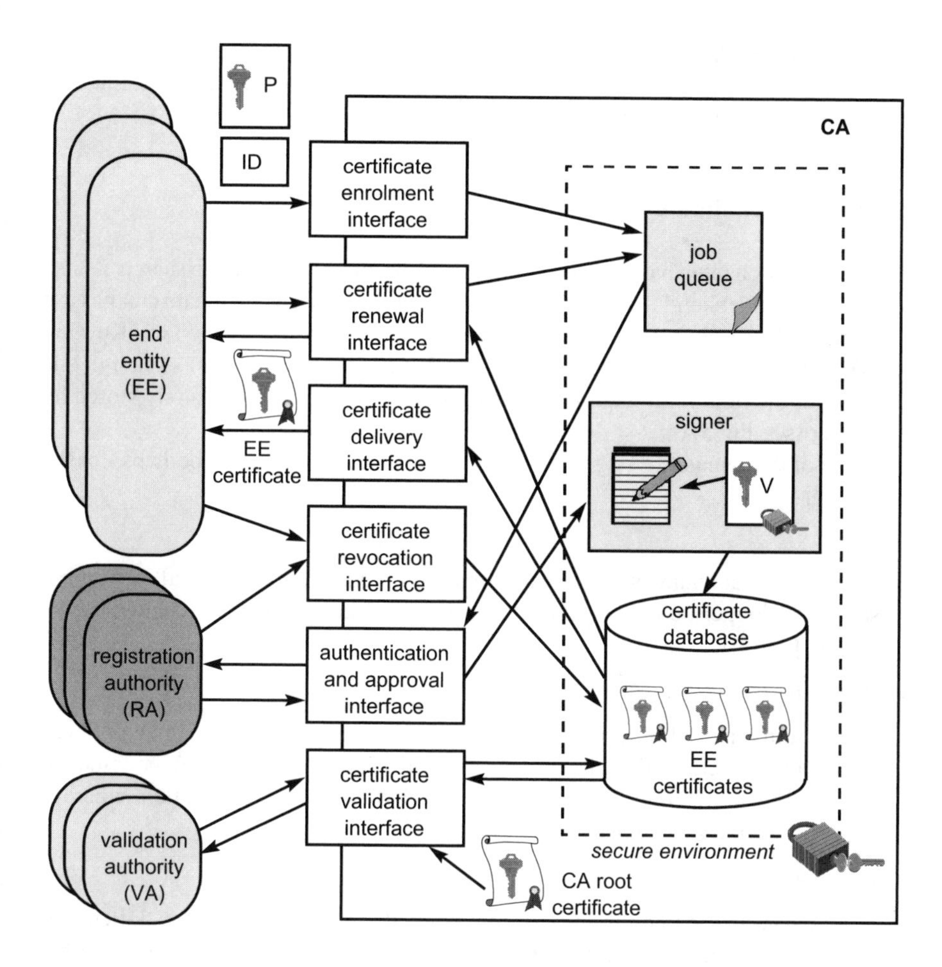

Fig 5.3 A complete picture of a generic CA with all the components identified.

It is the breadth of this last item that can be easily misjudged. An EE, so far, has been presented as a lone personal computer (PC) user with a browser (for enrolment) and, perhaps, an e-mail account (for delivery of an issued certificate, or a pointer to a certificate collection service). There are, however, many different types of EE in existence. Because a PKI can be used at any of the connectivity layers, different types of devices might be involved. For example, a firewall, or router, may be providing a tunnelled connection at the network layer in the form of a virtual private network (VPN). These devices are stand-alone and do not have 'users' like a PC, and may not have a Web browser either. Devices like this need to use a more direct interface to the CA and often submit a certificate signing request (CSR) straight to the CA over a programmatic connection.

This broad application of PKIs across different EE types and connectivity layers is a testimony to their versatility and only serves to emphasise that PKIs must be considered to be application, and layer, independent.

5.2.5 Certificate Life Cycles

Although this chapter has largely concentrated on the hierarchical issues it is also important to understand the life-cycle stages through which a certificate can pass. It all starts with the generation of a fresh public/private key-pair (called the KeyGen stage). Several options (i.e. as to who actually generates the keys) exist for this process but the minimum requirement is for the private key to be securely stored by the owner (see Fig 5.4).

Many of the remaining stages should be familiar by now, except, perhaps, for the following.

- Expiry

  A certificate contains, within its information fields, an expiry date after which it should no longer be trusted. Its value is included in the signature created by the issuing CA.

- Renewal

  For a limited period of time, a certificate that has nearly expired can be renewed — i.e. a fresh certificate, with a further period of active lifetime, can be issued to replace the dying one. This crossover period is the only situation where a duplicate identity can exist in the CDb.

- Revocation

  If a key is lost, or compromised, the associated certificate can be 'killed off' earlier than its natural expiry date. The CDb is updated to reflect this event but there are serious implications in doing this. The main problem is that most other entities will trust a certificate as long as it is offered within its expiry date. An extra source of information (about revoked certificates) is needed to support this condition called a certificate revocation list (CRL). CRLs are addressed later.

5.2.6 How a PKI is Used

Many different devices and software applications can use a PKI. For clarity, in this section, it is assumed that the party that is trying to establish trust is a Web browser — this is the 'client'. This client has its own certificate, EE1a, issued by CA1. The party offering its credentials to the client is a Web server — this is the 'server'. This server has its own certificate, EE2a, issued by the (different) authority CA2. The

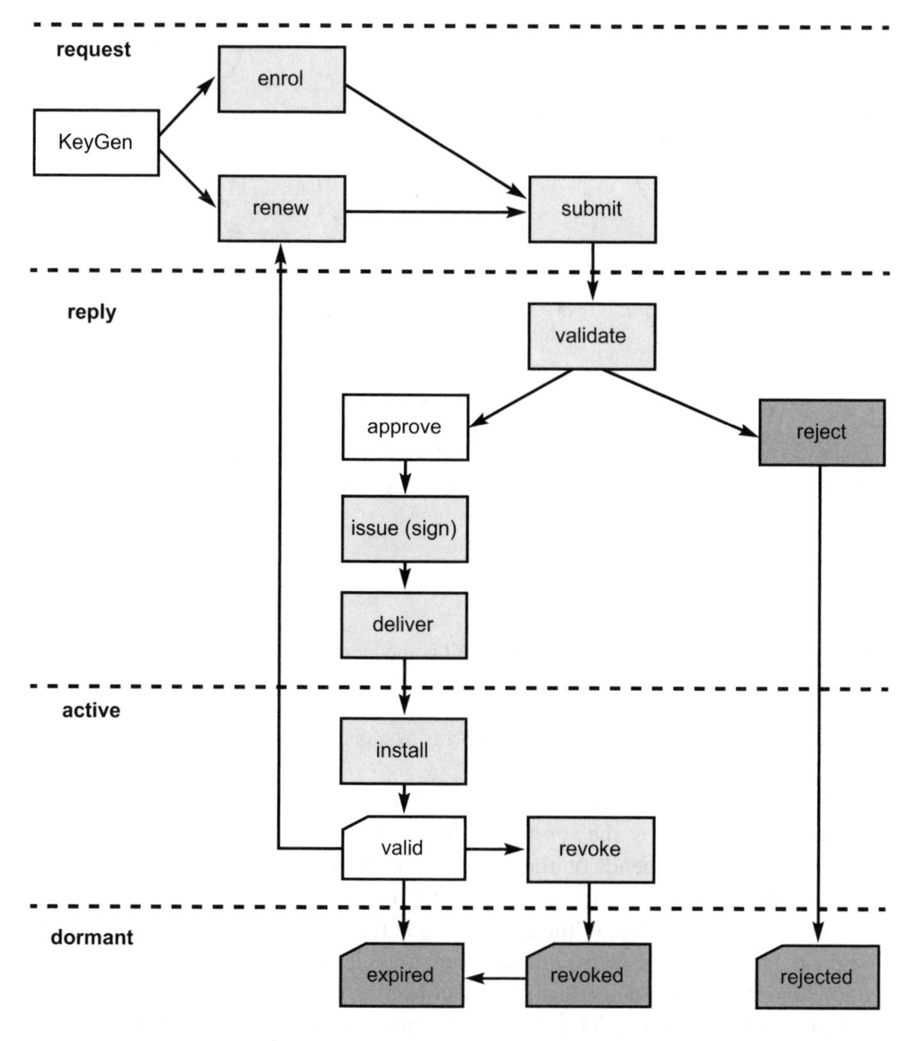

Fig 5.4 A typical certificate life cycle from cradle to grave.

transaction between the two parties is to be secured on the basis of the Web server's certificate being found to be trustworthy (see Fig 5.5).

Taking the viewpoint of the client, Fig 5.5 shows within the bold dashed line which PKI elements are needed for the client to check the server credentials.

This is the most popular client view that is implemented today and is an inverted POT. This allows the client to interwork with a number of PKIs from the perspective of being a single end-entity. The client supports a certificate trust list (CTL) in order to achieve this viewpoint and may 'trust' many external self-signed

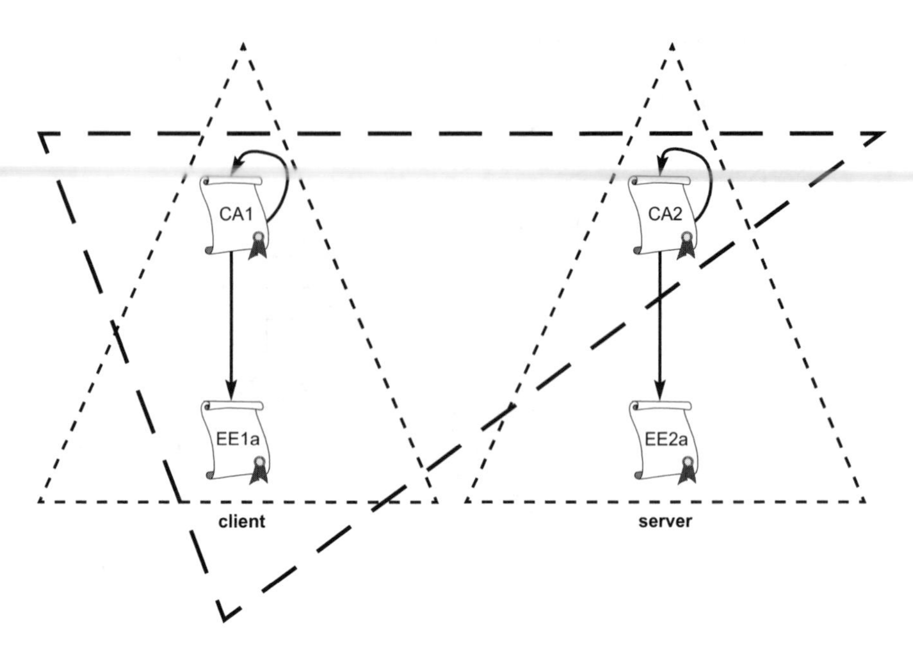

Fig 5.5 A client view of the necessary PKI components required to trust a certificate.

CA root certificates in this way; the problems associated with CTLs are covered in a later section.

The steps that need to be taken by the client to satisfactorily establish trust for the EE certificate offered by the server are shown in Fig 5.6 (the order in which the steps are performed depends on the amount of computation, speed of response, etc, and may vary slightly). Notice that CRLs are also mentioned to allow revocation-status decisions to be made (although any validation authority response-mechanism would do here and this is explained more fully later on). The steps show what should happen and do not necessarily reflect what actually happens in contemporary software implementations. The process becomes proportionately computationally longer if intermediate CA roots are present in the trust 'chain'.

The server certificate, EE2a, is sent to the client, for checking. If (and only if) the outcome of the checks is acceptable, the client can then process the actual data that was received along with the original certificate. For example, this might be validating a digital signature for the data block, or recovering an encryption key that was, itself, encrypted by the server's private key. The actual steps will be revisited in later sections, where different types of pyramids of trust are employed, and some of the implications for the client will then be explored.

It should be noted that before any message dialogue commences between the client and the server, it is also desirable that the client has performed a 'self-check' to validate its own certificate (EE1a). This confirms its own identity, and current

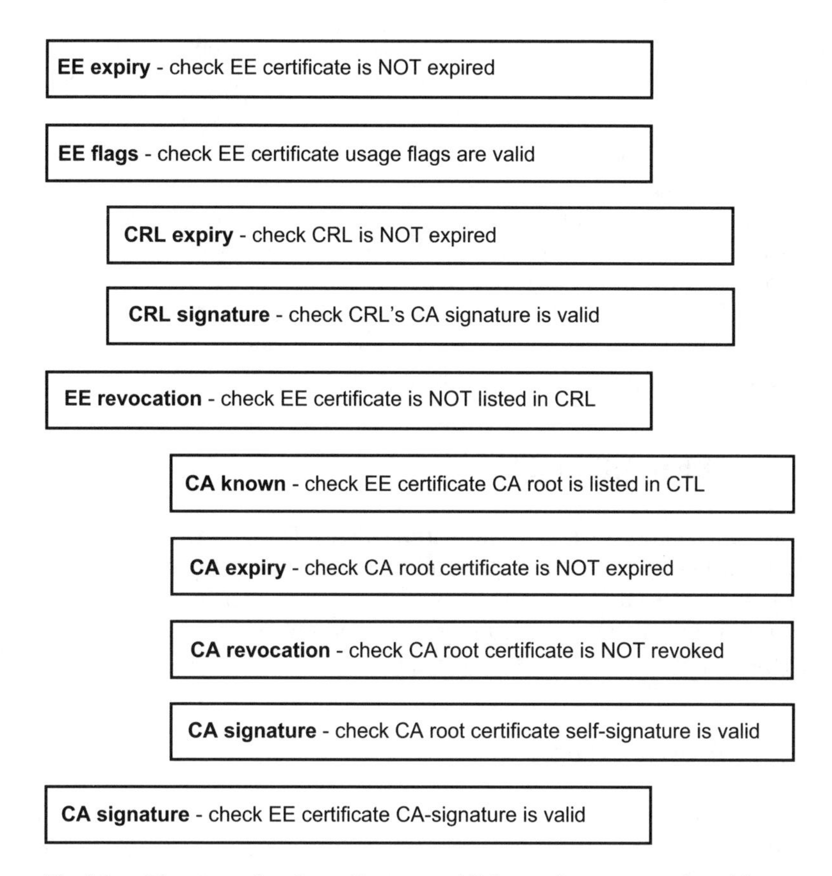

Fig 5.6 The steps taken by a client to establish trust in a presented certificate.

certification status, before it starts talking to other entities. Not many applications actually do this at the moment.

5.3 The Main Problem with a PKI is .....

Despite being widely recognised, and accepted, none of the following have a single universally accepted global solution:

- currency (despite the mighty dollar being widely recognised and accepted);
- language (despite English being widely spoken as either a first or second language);
- religion (despite having some similar principles, deities, and requirements for worship);

- plastic money (although most solutions are truly globally accepted, and formatted on a single, universally recognised card format, there is still no single solution because many large operators exist).

Given that these are all very significant widespread parts of everyday life, then why is it reasonable to expect a single, global PKI to exist? PKI vendors obviously want customers to believe that their particular plan for 'world domination' of the PKI space is the right one to select. But what about monopoly controls, legal restrictions and even technological preferences?

It is therefore assumed that the single global PKI will never happen.

Now that this statement has been made, and the single POT is off the radar screen, some further problems can be highlighted.

5.4 Other PKI Problems

There are other problem areas with current PKIs. These range from getting the certificate in the first place, using certificates correctly, maintaining trust within inverted POTs, legal applicability of interworking POTs, right through to application workload and responsibility. The first problem concerns the behaviour of the most popular user application to employ PKIs today — the Web browser.

5.4.1 Browser Wars

The Internet has developed a lot since Tim Berners-Lee, while working at CERN, invented the World Wide Web (WWW) in 1992. It was based on a text browser and was used for the sharing of academic papers on-line. Soon after (1993) Marc Andreessen, and a group of students from the National Center for Supercomputing Applications (NCSA) at the University of Illinois at Urbana-Champaign, released the 'Mosaic' browser for X-Windows. This was the first graphical user interface (GUI) browser and was planned to target all the common operating systems. In 1994 Andreessen left NCSA to help Mosaic Communications. In October of the same year Jim Clark (who had previously helped form Silicon Graphics) and Andreessen formed Netscape Communications. Netscape produced several new server products, but also rewrote the browser known as Mozilla (now Navigator and Communicator). These browsers were initially distributed, for free, to individuals and educational users. In 1995 Microsoft introduced Internet Explorer (IE) and the browser wars began.

During 1994 Netscape realised that they needed to add a security solution to their browser, to secure the various application protocols that the browser was to offer. Netscape approached RSA Security Inc who assigned George Parsons, director of Certificate Services, to the problem. Soon RSA, and Netscape, concluded that public key certificates, in conjunction with what has become known as the secure sockets layer (SSL) protocol, was the way forward.

To this end Kipp Hickman, at Netscape, created the first version of SSL (version 1.0) — this was never released publicly. RSA, and Netscape, concluded that there was a real need to establish a company capable of offering a service that established security, and trust, on the Internet. This company was named VeriSign Inc and George Parsons moved across from RSA to build an initial engineering solution. The company was launched in 1995. SSL gradually evolved and was eventually adopted by the IETF and renamed to the transport layer security (TLS) protocol (Fig 5.7).

Resulting from the need for certificates, a number of proprietary behaviours emerged between Netscape and Microsoft browsers. Unfortunately these differences mean that a CA service provider that offers Web-based enrolment, such as BT TrustWise, must provide two different sets of Web pages.

Additionally it is also necessary to support initial pages to identify the browser type, and version, to ensure that the enrolment process is adjusted correctly. Table 5.2 identifies some of the proprietary differences.

These inconsistencies mean that supporting users, and their browsers, is non-trivial. However, some of the issues have been fixed with later releases of the browser software. Persuading users to actually upgrade their browser is a difficult task but was simplified due to two fortuitous happenings.

The first was VeriSign's Root CA certificate expiry (this was hard-coded into older browsers) on December 31 1999 (which was an interesting day for IT systems in general). This encouraged many people to upgrade. The second is that more recently almost every browser has been revised due to potential security problems being uncovered by the user community. As a result both Microsoft and Netscape offer a 'Smart Update' solution. These 'enforced' revisions have also begun to address some of the proprietary behaviour of the browsers.

Looking back over the browser wars Netscape has gone from 80% market share to below 20% today. Indeed Netscape 5.0 (Gecko) did not get distributed beyond the 'developer' community, while Netscape 6 (which offers the greatest compliance to XHTML v1.0) has many other problems. Version 6.1 is expected to be the first mass-distributed release, and may determine whether it will survive. Microsoft IE beta 6.0, due for full release very soon, has replaced Netscape Navigator as the browser of choice for features and ease of use.

It has not been the authors' intention to ignore the many other excellent browser products out in the market-place — most mimic the main two products in some way (for compatibility reasons). Consequently they too can exhibit the same behaviour and also suffer from the same problems that are described above.

Attentive readers may be wondering what, if anything, became of Mosaic (the original NCSA browser). A clue is contained in the 'help — about Internet Explorer' menu in Microsoft's IE browser.

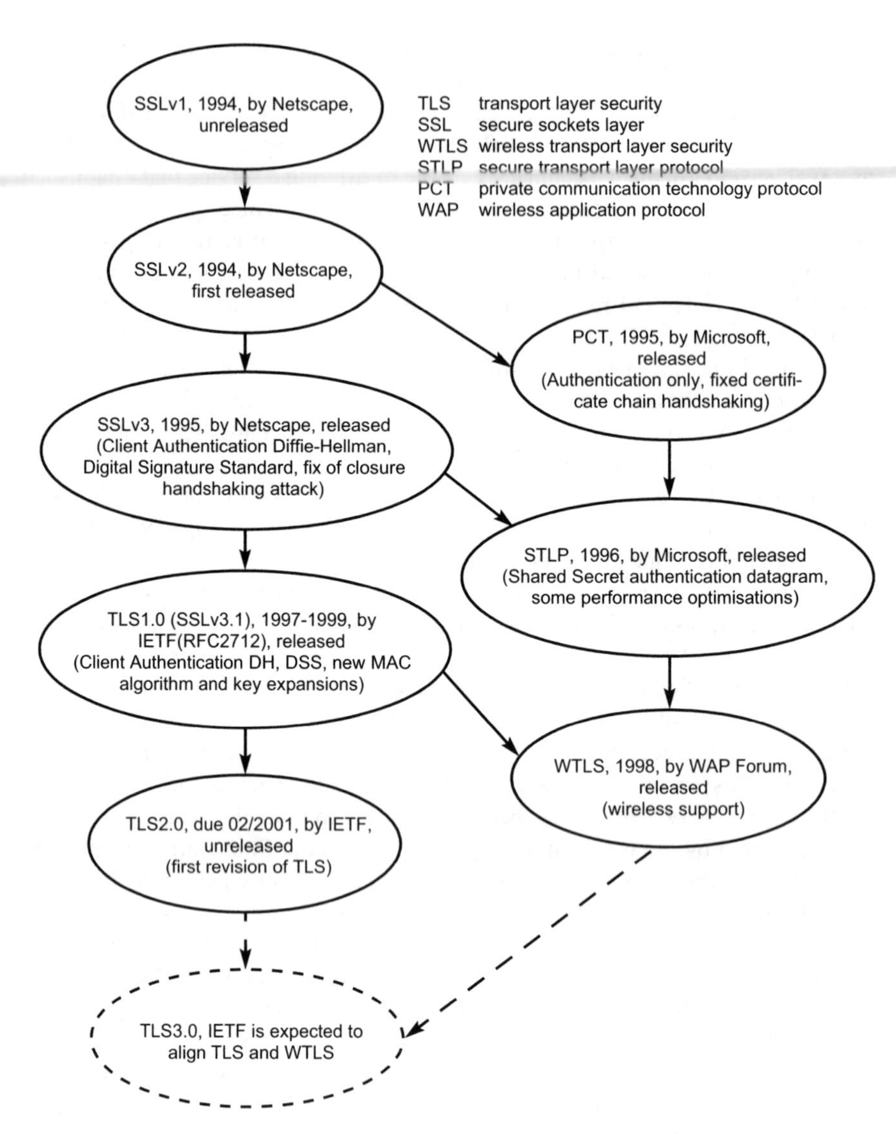

Fig 5.7 The evolution of SSL and other related protocols.

5.4.2 Certificate attributes

Attribute flags such as 'only to be used for signing e-mails' or 'only for user authentication' can be included in the body of a certificate. However, it is largely up to the software handling a certificate to judge whether the flag values are appropriate for what the user is trying to do. The flags have limited meaning and

Table 5.2 Some of the proprietary behaviours exhibited by Netscape and Microsoft browsers.

PKI functionality	Microsoft Internet Explorer	Netscape Navigator/Communicator
Key-pair generation	Uses CryptoAPI library v1.0 or v2.0	3.0+ uses browser KEYGEN function
Key storage	Centralised O/S key store	Local browser key store
Public key certificate export (excluding the private key)	.CER files — either distinguished encoding rule (DER) encoded X.509 **or** base-64 encoded X.509	Not implemented
Public key certificate export (including the private key)	.PFX files — personal information exchange (similar to PKCS#12)	.P12 files — PKCS#12 format
Public key certificate import (excluding the private key	.CER files — either distinguished encoding rule (DER) encoded X.509 **or** base-64 encoded X.509	.CER files — base-64 encoded X.509
Public key certificate import (including the private key)	• .PFX files — personal information exchange (similar to PKCS#12) • .P12 files — PKCS#12 format	• Limited ability to import PFX v2.0 • .P12 files — PKCS#12 format.

cannot be changed, or extended, after the EE certificate has been signed by the CA (otherwise the 'alteration' would invalidate the certificate signature). This limits their usefulness and a better solution might be to use a flag-less certificate that holds a pointer to a central user profile store. The profile can then be updated (in a controlled environment) in accordance with changing user roles and requirements.

5.4.3 Certificate Trust Lists

The technical use of CTLs has already been covered. However, current implementations normally allow users to manage CTL content for themselves. This means that, presented with a simple 'Do you want to trust this CA?' dialogue, the user can innocently choose to 'trust' anyone they like. Worse still, unless the user updates their software regularly, they will be frozen in time and will be using an increasingly out-dated CTL. Furthermore, there is only a limited ability to check the listed CTL root certificates (essentially by their expiry date — and CA root certificates usually have long lifetimes, perhaps 10 years) and little account is ever taken of possible revocation within this period. The CA private key might be compromised, or it may simply go out of business during this time.

CTLs are the equivalent of a CD player with one disc already loaded and the door glued shut. They are out of date before the user even gets to use them and cannot easily be changed in a controllable way.

5.4.4 Privacy PKIs

The current PKI model requires the user to submit their details to a CA for its RA to perform checks and, hopefully, approve the application. The resultant certificate normally contains some user-specific information that is bound to the user's public key. It is increasingly evident that there is a market for 'anonymised' EE certificates. Producing EE certificates within such a privacy PKI is done by performing the normal checks, and then including a unique identifier code in the certificate instead of user-specific information. Since the EE possesses its private key (that matches the public key held in the certificate) their rightful proof of ownership of the key-pair is never in dispute, but the privacy PKI ensures that their user-specific details are not publicly on display in their certificate. The CA can verify the material facts for an enquirer.

Thus, questions can be asked like: 'This EE certificate was presented to me by someone claiming, in an associated piece of information, to be Joe Soap — is this the identity of the rightful owner?' and: 'This EE certificate was presented by someone claiming their postal address is 1 The Street, Anytown — is this true?' The enquirer has to gather the specific extra information it needs from the user and then correlate it via a response from the signing CA (who has the definitive information from the original user enrolment).

This has confidentiality benefits for users as they do not have to show all of their details when presenting their certificate to a complete stranger. Better still, they can enlist the help of their CA to report on the details being requested by enquirers and possibly restrict, or prevent, responses where necessary.

Specialised services like this are not widely implemented at present. In many respects, they go well beyond the traditional role of the CA. There is, however, a clear need, and many benefits, for this type of PKI and it will have to be addressed in the future.

5.4.5 Legal Jurisdictions and Trust Anchors

Several world economies have now implemented electronic signature laws. Some give legal effect to accredited CAs. As a consequence, the issue of how a sender, and a recipient, know what legal standing is offered by a certificate must be addressed.

Within the European Union the 1999 European Directive on a Community Framework for Electronic Signatures [1] has established the high-level principles on legal recognition, national supervision and also a voluntary framework for the accreditation of service providers. The European Electronic Signature Standardisation Initiative (EESSI) [2] is taking the work of the parliamentary draftsman and producing technical specifications which engineers can understand and implement.

Within the UK, the Electronic Communications Act, 2000 [3] (ECA) was principally concerned with implementing the requirements of the Directive into UK law. The ECA was the end-point of a protracted policy debate, the full extent of which is beyond the scope of this chapter. Suffice to say that in mid 1999, following some two years of intensive lobbying by the private sector, Government requested detailed proposals for an industry led body to regulate trust service providers and tScheme [4] was conceived.

tScheme was established as a limited company, an interim board was formed, and it was clearly positioned as the UK Government's preferred conduit for the voluntary approval of trust service providers. Since then a methodology for developing approval profiles (the criteria to be adopted) has been created and a process for external auditing devised. There are currently two pilot evaluations under way. More profiles are under development with priority given to those for certificate services.

In future, it is less likely that a recipient will have an existing relationship with a sender's issuing CA and therefore the recipient must ensure that the sender is presenting the certificate for use in the context for which it was intended. This context will be defined within the issuing CA's CP and CPS documents but these are lengthy and are almost impossible for a machine to interpret reliably (though some work has been done to explore solutions to this problem). Indeed, although they are readable by humans (just!), CPs and CPSs are often not easily understood by a layman.

In future (and at the risk of making certificates significantly larger in size), additional certificate extensions might include issuer (and EE) extensions that define the legal environment(s) under which the issuer CA (and EE) might be trusted.

Moves towards legal certainty are presented in such a way that they appear to improve the poor security that is, apparently, preventing the take-up of eCommerce. This is certainly true within the consumer and SME space as surveys repeatedly show [5]. However, business people are prepared to accept a higher level of risk than consumers do, since they see any losses as a part of the 'Cost of Sales'. Nevertheless a recent survey by the Delphi Group indicates that companies view trust as vital to collaborative commerce [6]. Techniques are therefore necessary to foster the deployment of a well-founded risk management model on-line, in the same way that large organisations do formally, and smaller ones do more intuitively, in the physical world.

5.4.6 The Managed Service Provider

Within a managed service provider (MSP) environment, the EE certificate is just a small part of a larger product and will not be seen, or handled directly by the actual end user at all. Many PKI systems do not cater for this operational model. Serious problems occur when enrolling for certificates (usually a bulk-enrolment is required with associated bulk key-pair generation and workflow elements for integration with existing provisioning processes) and when revoking certificates (this should only be a function that the MSP administration can perform as part of the product management processes). It is very difficult, today, for an MSP to use a PKI (even if it can be technically achieved) in an economical manner.

5.4.7 Legacy Issues

Whenever a piece of information is 'signed', its future must be qualified. If it is to retain its value for a long period of time (and not all information needs this feature) it must be possible to archive the information and revalidate its integrity at a future point in time. Typically, a CA will state, in its CPS, that it will retain records of all certificates it has issued for, say, 30 years and will guarantee uniqueness for that period of time. This only goes a small way to facilitating the revalidation process though. It is a brave assumption that the cryptographic strength of the algorithm used to protect the information has not become trivial over time (and hence the signature susceptible to forgery). Similarly, to presume that the information is still stable, and available on a media format that can be read by future hardware, might be unrealistic too.

The legacy of current information will primarily depend upon the continuity of record-keeping. For example, today, if an individual creates a will and lodges it with a solicitor for safe-keeping, it is a matter of professional continuity that, if that firm ceases to trade, its archives will be passed on to another law firm to preserve the availability of the paperwork.

With the speed of progress in the computing world, the PKI legacy problem may well become equivalent to a modern day lawyer having to maintain, and still be able to read, wills carved in hieroglyphics on tablets of stone that are damaged and degraded. Furthermore, being able to read what is left of the symbols may not prove their meaning either — the determination of context, and resolution of ambiguities, can make the material vague, and of insufficient enforceable legal standing, in a modern setting.

5.5 Current Solutions to Globalisation

It has taken a large part of this chapter to present the position of PKI technology as it is today. This has been because, as with any journey plan, it is necessary to know both where you are setting off from **and** where you actually want to end up going to. The old adage about the reply you might get when asking directions being: 'If I was going to go there, I wouldn't set off from here,' is highly applicable to present day PKI implementations.

Not all of the problems outlined in the previous section actually have solutions today. For example, the browser wars are equivalent to car manufacturers deciding their own layouts for pedals and switches (how many times do people try to indicate for a left turn but end up wiping the windscreen instead?). This does not make a car less drivable *per se* (although varying the order of the foot pedals might!), but can make the user experience far more challenging. The hidden internal workings of cars, or software, are largely the choice of the manufacturer but the interfaces are natural candidates for standardisation wherever possible. Interestingly, the same browser battles are now being fought in the mobile telephone arena with WAP micro-browsers exhibiting similar proprietary behaviour.

Next generation PKIs will have to address these problems but, at least for now, suggestions and solutions do exist for some of the globalisation issues.

5.5.1 Cross-Certification

In a world where multiple CAs, and POTs, are all trying to interwork, one proposal is to modify the client's inverted POT. This is done by getting each pair of interworking CA operators to have agreements and then to cross-certify their CA root certificates (see Fig 5.8). A single local CA root (trusted by the client) can then be used to enable trust for the other CAs without having to hold all the other roots in the local CTL. An intermediate cross-certified CA root has the same identity (and key-pair) as its self-signed equivalent. It can, therefore, 'logically and locally'

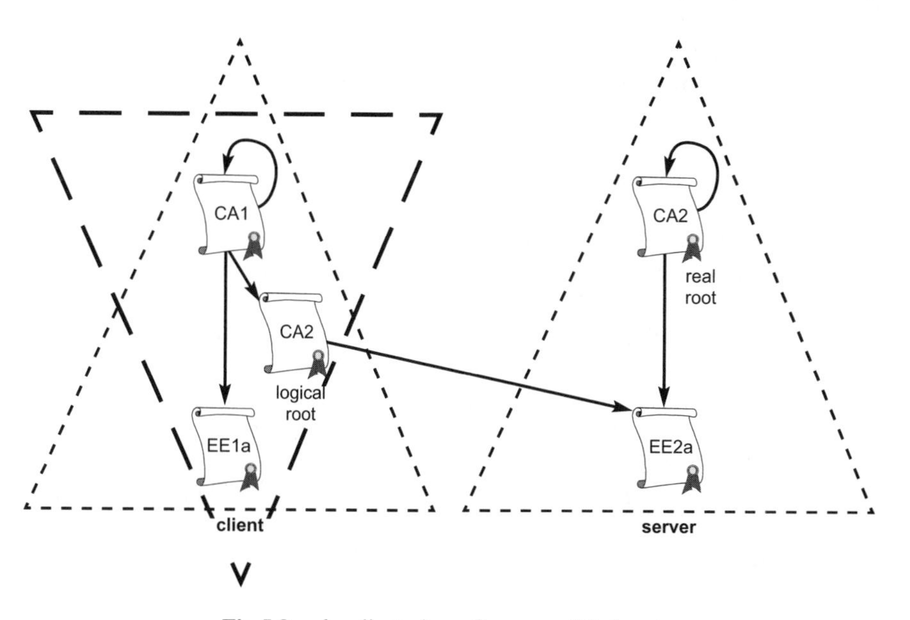

Fig 5.8 he client view of cross-certified roots.

replace the 'real' self-signed CA root. This still requires that the intermediate CA root is either stored locally by the client (in an intermediate CA's CTL) or that it is sent with the EE certificate each time it is presented for checking.

There are flaws with the solution described above and depicted in Fig 5.8:

- this would require each CA owner to agree to interwork with the other CA owners at both a business and a technical level — national legal frameworks, and purely commercial differences, could have a severe impact on the practicality of such agreements;
- the number of cross-certifications necessary to, potentially, enable any client to trust all the other CAs (say 'n' CAs in total) is $n(n-1)/2$ — mutual, bi-directional cross-certification must be treated as two separate agreements (at least from the root certificate signing perspective), effectively doubling this number of interconnections;
- the workload for the client increases as all non-local PKI EE certificates present a longer chain for validation — this extends into CRL checking both for EE and CA certificates too;
- client complexity is also increased because the ability to handle larger CTLs and/or more complex inbound certification material is required.

Cross-certification can help with small, localised trust agreements where, say, an employer issues certificates for all of its employees at a particular time and then moves to a different CA provider (or perhaps undergoes a merger or acquisition). Corporate interworking is required (at least for a year or so) to avoid having to rapidly replace all the old certificates with ones from the new CA. Beyond this scenario, cross-certification cannot provide a truly global solution.

5.5.2 Bridge CAs

A solution favoured by some national organisations, for large-scale interworking, is to create a super-CA root above all others. This super-root is used to certify all other CA roots and is trusted by all clients. There should be a slight problem becoming visible here. All that this achieves is to agree that a single PKI will never happen and, in light of this, to go and create one anyway. Effectively this just makes one global POT big enough to absorb all of the smaller POTs underneath it.

The normal self-signed CA root certificates are discarded and replaced by a CA root certificate signed by the bridge CA and a self-signed bridge CA root certificate (Fig 5.9). Clients are not only required to check longer chains, but also have to be able to locally select the peer CAs that they wish to trust from a standard 'approved CA list' provided by the bridge CA itself.

The bridge CA creates a super-hierarchy and consequently suffers from the same problem as a local hierarchy — if a client needs to trust an outsider, either the

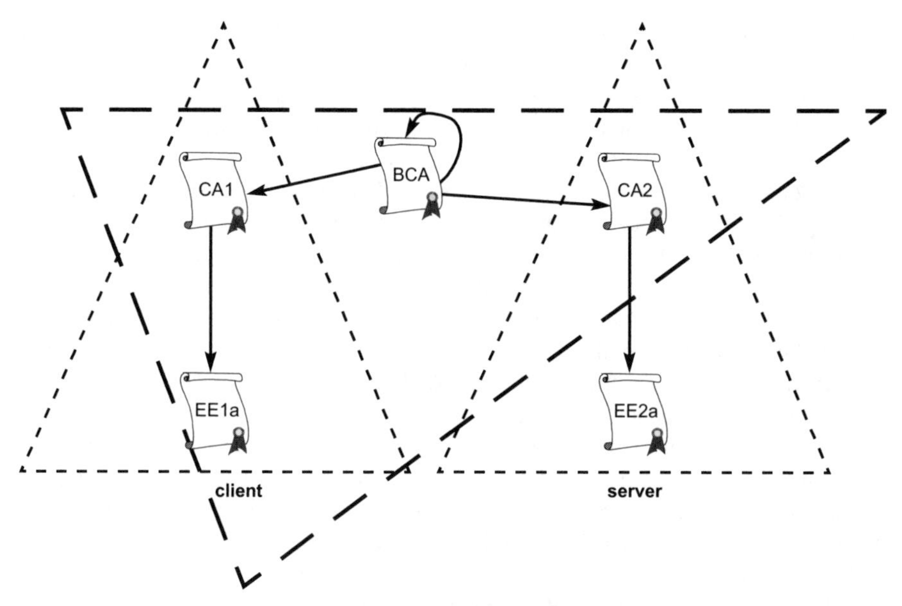

Fig 5.9 A typical bridge CA hierarchy.

outsider must join the bridge community, or the client must separately add the outsider's self-signed root alongside the bridge CA root in its CTL. Determination of trust (via the CTL contents) thus remains with the client.

5.6 Certificate Validation

Figure 5.3 mentioned a 'validation' interface for a CA. This is usually called the validation authority (VA). The format of a VA ranges from a simple repetitive query running against the CDb to find revoked certificates (and present them, via a suitable download protocol, as a signed CRL), through to more sophisticated real-time query-response mechanisms like on-line certificate status protocol (OCSP) and simple certificate validation protocol (SCVP).

5.6.1 Certificate Revocation Lists

Today, a validation service is not much different to that which was originally offered by a credit card company whereby a list of stolen, or cancelled, cards was distributed to merchants on a periodic basis. After a CA has issued a certificate to a trusted individual that individual can use this on-line electronic identity how they choose. It is the task of the recipient (the relying party) to confirm that the user's trusted status has not been revoked by the CA. This relationship is often called the relying party agreement, where relying parties are required to establish the ongoing validity of presented identities for themselves.

In the same way as for a credit card 'black-list', an issuing CA regularly constructs a CRL that identifies its revoked certificates 'that must not be trusted' any more (despite them still having valid expiry dates, attributes and signatures). A CRL is dated, and then signed, by the issuing CA to show its authenticity and is issued perhaps on an hourly, or a daily, basis. It also has a defined validity period.

To use a CRL, a client must first obtain it from the issuer. There may be several CAs that a client needs to trust and there will be a CRL for each one. Each EE certificate issued by a given CA should contain, within its signed information fields, an absolute reference to the CRL resource for that CA. It is a matter of risk management to allow for the fact that revoked certificates cannot appear on the list until the next CRL update is created by the CA.

Because the client uses the logical decision 'if a certificate is not listed in the CRL, then it is OK' as part of the checking process, a revoked certificate will be trusted until it appears in the list. In fact, the only correct logical inference that can be made is 'if a certificate is listed in the CRL then it is not OK'.

Further to this basic flaw, two major issues surround CRLs and their use.

What happens if a fresh CRL cannot be obtained (and the previous one has expired)?

Making the validity period much longer than the update period allows for any gaps in service, but worsens the potential at-risk time during which new revocations are not yet listed.

What happens to the audit trail when a revoked certificate reaches its natural expiry date and drops off the CRL (as happens with many current CA implementations in order to limit the size of the CRL)?

A 12-month certificate that is revoked after 6 months will produce signatures that appear invalid for the remaining 6 months before its expiry (as far as the client is concerned). However, these will suddenly appear to have been perfectly valid after the expiry date (when they are archived and examined at some future time) because the CRL ceases to mention them.

5.6.2 On-Line Certificate Status Protocol

OCSP allows certificate status to be validated in real time. There is no waiting for the next CRL to be published, and no double-negative logic contributing to the client decision. It is a little heavy on the client because the protocol only handles single certificate queries. This means that checking a trust chain requires the client to possess all the participating certificates and to present them one at a time to the OCSP responder. The replies are basically limited to 'yes' or 'no'. This very much echoes the real-time credit card validation where a merchant swipes a card through a small terminal to locate and reserve funds for a transaction.

The underlying technology may be connecting directly to a CA's CDb, or may simply be creating a virtual on-line response based upon a standard CRL (but saving the clients from having to download entire CRLs at every update interval). It should be noted that an OCSP responder must be available on a 24 × 7 basis.

5.6.3 Simple Certificate Validation Protocol

SCVP works in a similar way to OCSP, but allows the client to offload much of the certificate chain checking work to the SCVP responder — only a 'starting point' certificate needs to be submitted by the client. Replies can be more than just a simple 'yes' or 'no', but the responder must still always be available.

Today, SCVP responders are very new (and rare) and OCSP responders are only gradually beginning to appear. They are often provided by a CA (as the VA component) and consequently do not offer assistance for multi-POT hierarchies operated by different CAs. They do, however, go a long way towards solving the problems associated with CRLs.

5.7 Better Solutions for the Future

To overcome the shortcomings mentioned in the previous section, centralised resources may offer an answer.

5.7.1 Centralised policy

The concept of client-based CTLs is difficult to manage effectively. One way to improve this (at the expense of increased network traffic) is to provide a centralised CTL store. Client applications can then refer to this to see if they should 'trust' a root CA certificate (and hence the EE certificates issued by that CA). Such a centralised resource might be provided (and updated) by a browser manufacturer, or even by a corporate intranet operator. This scheme can also provide the intermediate CA certificates in the case of partitioned CA hierarchies.

The trust policy can, therefore, be set to suit the trust domains that are allowed. Policy can be base-lined on a generic public policy, and modified by a localised private policy. Updating expired CA root certificates (and even removing revoked ones entirely from the CTL) is nicely co-ordinated and the need for CA root certificate revocation status checking is largely unnecessary.

Clients can reduce their local trust list to a single CTL provider's root certificate with which all CTL responses are signed for authenticity. However, the workload is still very much upon the client to do the actual establishment of trust. Protocols to access centralised CTLs and client support are not widely available today.

5.7.2 The Validation Broker

A validation broker (VB), as its name implies, is able to provide validation information beyond the local scope of any one particular CA. It may be tied to a CA or be entirely independent.

The job of a VB is either to answer, or else to refer onwards, client queries about certificate status (see Fig 5.10). Clients are configured with a primary VB resource location. Responses can apply to a range of CAs without ever needing cross-certification or bridges. The VB has agreements with its participating CAs (and hence such relationships are centrally managed and trusted) and can offer responder interfaces for any of the current, or future, status-checking protocols. Its information can be sourced from CRLs, or direct CA OCSP responders, or from local information databases. All responses are signed by the VB and are audited. Signing proves the authenticity of a response, and auditing gets around the problem of lost CRL information (after a revoked certificate's expiry date).

The VB can further enhance its value-add by offering policy-based replies during exceptional circumstances (e.g. for covering periods when a CRL has expired, or an OCSP responder is off-line). If a VB is asked about a CA with which it does not have an existing agreement, it can either defer until an additional agreement exists, or refer on to another VB (that it knows about) that does have an existing CA agreement. The overall reply can come from either VB and can be trusted because of the provable VB-VB relationship.

The VB can also automatically refer to a neighbouring VB if it has to be off-line itself. Ultimately, a client can be simplified to contain only its own certified client identity and a single validation broker 'root' certificate that contains the VB public key and VB resource location. All client-originated certificate-validation requests (including self-checks) can be sent to the VB.

Validation broker technology, and services, are currently being developed. Very recent IETF standards address delegated path validation (DPV) and delegated path discovery (DPD) protocols. Some commercial products are available to perform basic VB functionality mentioned in this section; however, just like the early PKI days, VB products and services are currently heading towards their 'gold-rush' era and need to be carefully studied, and analysed, before they are selected for use.

5.8 Summary

This chapter has presented the history, to date, of the PKI and has also shown some of the proposed ways forward for its evolution. Just as the petrol engine reached a point where carburettor fuel systems could deliver no further improvements in performance, economy, and emission controls, PKIs also seem to have reached their current limits (essentially due to the use of 'off-line' validation mechanisms).

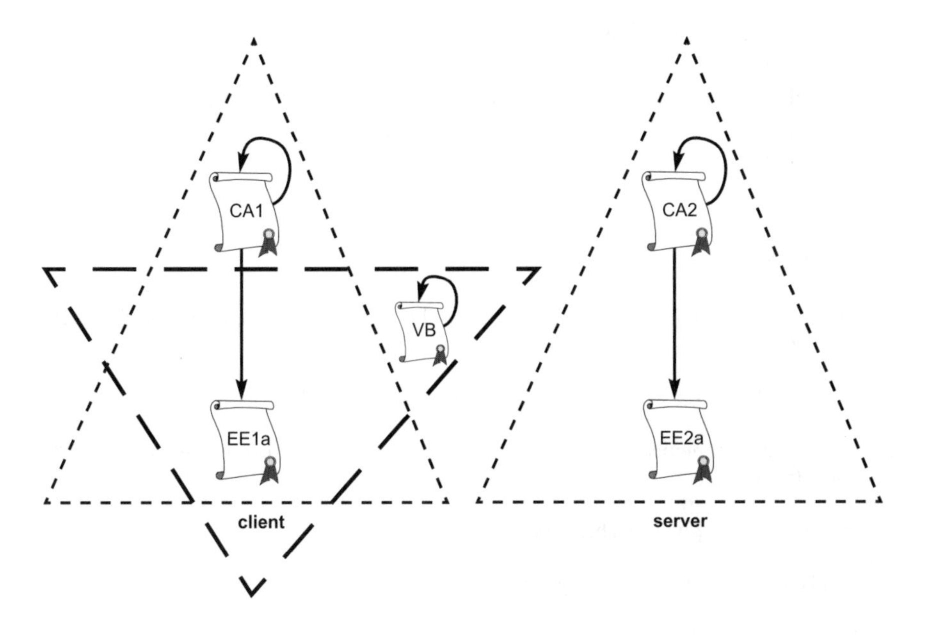

Fig 5.10 A validation brokering service.

Fuel injection systems and electronic management provided a new lease of life for engines. Validation broker services (and OCSP/SCVP) will do the same for PKIs.

Today, PKIs are very good when used in a localised context, but are not well suited for use in a global and (ever more rapidly growing) commercial context. It is clear that the most likely way forward for PKIs is to provide intelligent, enhanced, on-line validation components to allow simplification of the client and to achieve the necessary degrees of interoperability required for global commercial implementations.

International efforts for legal recognition of digital signatures need to succeed in parallel with this, but, it must be recognised that users crave usability. Technologists must engineer solutions that allow users to make informed risk management decisions even when this appears to be at the expense of absolute factual certainty.

Many of the techniques, and solutions, for interworking still have a place in this exciting future, but not necessarily in the location where they are currently portrayed. Although the bandwidth requirements may increase in this situation, the basic premise of being on-line in the first place (to obtain information or perform transactions) readily supports comprehensive on-line validation services. Arguments exist for both increased local processing (given the vast amount of client computing power now on the planet) and for more centralised network resources (given the better level of trust that can be managed from a single point of reference). As always the future will probably be a chaotic mix of the two.

The solutions required for next generation PKIs are far from easy to define, or implement, at present, but their evolution is definitely something to watch closely over the next few years.

References

1 EU Directive on Electronic Signatures — http://europa.eu.int/eur-lex/en/lif/dat/1999/en_399L0093.htm

2 EESSI — htpp://www.ict.etsi.org/eessi/EESSI-homepage.htm

3 Electronic Communications Act — http://www.legislation.hmso. gov.uk/acts/acts2000/20000007.htm

4 tScheme — http://www.tscheme.org/

5 Red Herring — http://redherring.com/

6 Delphi Group — http://www.delphigroup.com/

Bibliography

Adams, C., Lloyd, S., and Kent, S.: '*Understanding the public key infrastructure*', New Riders Publishing (1999).

Austin, T.: '*PKI — A Wiley tech brief*', Wiley (2000).

Feghhi, J. Williams, P. and Feghhi, J.: '*Digital certificates*', Addison-Wesley (1998).

Ford, W. and Baum, M. S.: '*Secure electronic commerce*', (2nd Edition) Prentice Hall (2001).

Grant, G. L.: '*Understanding digital signatures*', CommerceNet Press (2000).

Rescorla, E.: '*SSL and TLS — Designing and building secure systems*', Addison-Wesley Professional (2000).

Schneier, B.: '*Applied cryptography*', (2nd Edition) Wiley (1995).

Thomas, S. A.: '*SSL and TLS Essentials*', Wiley (2000).

6

AN OVERVIEW OF IDENTIFIER-BASED PUBLIC KEY CRYPTOGRAPHY

I Levy

6.1 Introduction

All public key cryptography (PKC) methods require, in some sense, a public and private key pair. It is the semantics of, and relationship between, these keys that determine the security model for a given public key cryptosystem. Shamir [1] proposed the concept of an identity-based public key cryptosystem (IDPKC) in 1984. Some years later, Cliff Cocks of the Communications-Electronics Security Group (CESG) invented the first practical solution to this problem. This chapter will explore the current model for PKC and examine the salient points differentiating traditional PKC from IDPKC, before then giving a broad overview of the Cocks IDPKC method and finally exploring some potential practical applications where IDPKC may give a real business benefit. IDPKC will be referred to as identifier-based PKC since it is a more appropriate nomenclature — the identifiers need not in any way refer to an 'identity' in the traditional sense.

All examples in this document are purely fictional and no parity with real systems should be drawn. Throughout this document, A and B will represent the two communicating parties. It is assumed that private keys are afforded the necessary protection commensurate with the sensitivity of the data they are to protect. In any public key cryptosystem, if this assumption is not made, then all trust and integrity statements can be dismissed. The further assumption that the authorities protect their secrets well is also required. In traditional PKC, compromise of a root authority's private key is sufficient to negate all trust in all certificates signed by that authority. In identifier-based PKC, compromise of any one of the authorities' private data sets is serious, but not fatal. However, compromise of all secret data held by the disparate authorities renders the system insecure.

6.2 Traditional PKC versus Identifier-Based PKC

6.2.1 The Traditional Process

The conventional view of PKC has the communicating parties responsible for the generation of their own key pairs. Suppose that A wishes to send B a secure message. Before this can occur, B must register a public key and publish it in an accessible directory. In order to accomplish this, B generates a random private key, derives the relevant public key from it and sends this public key to an authority. The authority, which must be trusted by all potential parties to the communication, validates the identity of B and then embeds this identification information, along with the supplied public key, into a certificate which is then signed using the authority's private keys. B (or the authority) then publishes the certificate in a publicly accessible directory. Now assume that the above has taken place and A wishes to communicate with B. A must firstly determine which public directory the certificate for B resides in. Then, the certificate must be retrieved and a validity and integrity check performed. Finally, A can extract the salient parameters pertinent to B, encrypt the communication and send it to B. A is assured that only the B specified in the certificate can read the message[1]. Upon receipt of the message, B uses his private key to decrypt the message. This proves that he was the recipient detailed in the certificate (since his private key can decrypt the message, his public key must have been used to encrypt it), but implies nothing about the sender. In order to gain assurance about the identity of the originator of the message, A must sign the message using a private key. Obviously, B must verify this signature on receipt and determine the validity of the purported originator. In order to do so, B must obtain the relevant public keys for A's signature. These are contained in A's certificate and so B must obtain, and perform an integrity and validity check on this certificate before extracting the salient parameters and validating the message source. As a side issue, B must also trust the certification authority that issued A's certificate.

In contrast to traditional PKC, public keys in an identifier-based PKC system are predetermined, perhaps by a conventional identity, e.g. name or e-mail address. Since the public keys are predetermined, it is unnecessary to retrieve, integrity check or validate them. Since there is no requirement to validate public keys, there is no need for recipient certificates and the associated directories[2]. The concept of 'registration' in identifier-based PKC is stronger than that of traditional PKC. In identifier-based PKC, private keys are derived from public keys using some secret information. Thus, registration involves an authority determining and supplying a user's private key based on their public key. In the naive case, the authority can read any user's communication (since it can implicitly derive any private keys

[1] Whether or not this was his intended 'B' is another matter!

[2] For the purposes of this high-level description, we shall ignore revocation.

necessary). This is countered by splitting the authority into two or more constituent parts. All authority constituents are required to create a private key. To emphasise the point that the logical authority is physically split into a number of constituent parts, it will forthwith be referred to as the (aggregate) authority.

6.2.2 The Identifier-Based Process

Consider the now-familiar situation where A wishes to send a message to B in a non-interactive manner. A can determine B's public key from his identity (which is generally known *a priori* in a standard format, for example an e-mail address), encrypt the message and send it. Explicitly, there is no need for A to obtain any further information about B before sending a message. Upon receipt of the message, B uses his private key to decrypt the message and can read it. This proves (modulo the identity verification performed by the aggregate authority) that he was the intended recipient (since his private key can decrypt the message, his public key must have been used to encrypt it), but implies nothing about the sender. In order to gain assurance about the identity of the originator of the message, A must sign the message using a private key. Obviously, B must verify this signature on receipt and determine the validity of the purported originator. In stark contrast to traditional PKI, no extra information is required to validate the originator's signature (assuming an identity-based signature scheme is employed). Validation of the signature requires only knowledge of the sender's identity (which is implicit from the received message) and the message itself. At no point is access to a directory required if we ignore (temporarily) the revocation issue.

Now consider the situation where A wishes to send a message to B and B has not registered with the (aggregate) authority. In a traditional PKC model, this is impossible since B must have generated his private key in order to create his public key which is then published by the authority. In the IDPKC model, if B has not registered, he simply cannot decrypt the message he is sent by A. A can still send the message since the public key is extant and predetermined. B may contact the appropriate (aggregate) authority after message receipt to obtain his private key and thence decrypt the message.

As with traditional PKC, both parties must trust the authority to deliver private keys only to suitable recipients. Furthermore, each communicating party must trust the counterpart to have kept secure their private key.

6.3 The Cocks IDPKC Method

This section presents a brief overview of the mathematics behind the Cocks IDPKC method. For a more complete description, refer to the papers on the CESG Web site [2].

6.3.1 System Set-up

For simplicity, the authority in this exposition will be a single entity — for details on how to split the authority knowledge securely, see the papers on the CESG Web site [2]:

- the authority begins by constructing a system modulus consisting of two primes, P and Q with $P = Q = 3 \bmod 4$;
- the authority makes available the system modulus $N = PQ$ and a hash function H such that:

$$\left(\frac{H(ID)}{N}\right) = +1$$

  where $(\div)$ is the Jacobi symbol and ID is an arbitrary string representing the chosen identifier. In practice, H is likely to be a standard cryptographic hash which is iterated until the required condition is true.

6.3.2 User Registration

The practical issues associated with registration (i.e. the act of proving one's identity to a third party) will be ignored; merely the mechanics will be dealt with here. Suppose a user's chosen identifier is ID. Then his public key is $a = H(ID)$ where $H(ID)$ satisfies the condition in section 6.3.1. The authority provides the user with his private key b satisfying either:

$$b^2 = a \bmod N$$

or

$$b^2 = -a \bmod N$$

Explicitly, the private key is a square root (mod N) of the user's public key (which is his hashed identifier).

6.3.3 Sending

Suppose Alice wishes to send Bob a message. She will send each bit of the message cryptovariable separately to Bob. For each bit of the message cryptovariable, she sets $x = \pm 1$, representing the sense of the current cryptovariable bit.

For each bit of the cryptovariable, Alice will:

- pick a random t such that $\left(\frac{t}{N}\right) = x$;
- send $s = t + \frac{a}{t} \bmod N$ to Bob;
- she must also send $s = t + \frac{a}{t} \bmod N$ (for a fresh t) since she does not know whether Bob's private key is b such that $b^2 = a \bmod N$ or $b^2 = -a \bmod N$.

6.3.4 Receipt

We assume that Bob's private key b is such that $b^2 = a \bmod N$ and he uses the $s = t + \frac{a}{t}$. The converse case is obvious.

To recover the original cryptovariable bit, Bob calculates:

$$
\begin{aligned}
\frac{s+2b}{N} &= \left(\frac{t + \frac{a}{t} + 2b}{N}\right) \\
&= \left(\frac{t + \frac{b^2}{t} + 2b}{N}\right) \\
&= \left(\frac{t(1 + \frac{b}{t})^2}{N}\right) \\
&= \left(\frac{t}{N}\right)\left(\frac{1 + \frac{b}{t}}{N}\right)^2 \\
&= \left(\frac{t}{N}\right) \\
&= x
\end{aligned}
$$

6.3.5 Security Analysis

We shall present two arguments concerning the security of this algorithm. Again, for a fuller discussion, see the papers on the CESG Web site [2].

6.3.5.1 Passive Attack

Given an attacker listening to the wire, can he recover x from the s values he will see without knowing b or the factorisation of N? Suppose that there is a procedure that recovers x from s without knowing b or the factorisation of N. Then, there exists a mapping:

$$F(N, a, s) \rightarrow x = \left(\frac{t}{N}\right)$$

whenever $s = t + \frac{a}{t} \bmod N$ for some t.

Now, consider the value of F when evaluated for an a where the Jacobi symbol $\frac{a}{N} = +1$ but a is not a square mod N. In this case Jacobi symbols $\left(\frac{a}{P}\right) = \left(\frac{a}{Q}\right) = -1$ where $N = PQ$. If t was the value used to calculate s, there will be three other values t_1, t_2, t_3 giving the same value of s defined by the simultaneous congruences:

$$\begin{aligned} t_1 &= t \bmod P & t_1 &= \tfrac{a}{t} \bmod Q \\ t_2 &= \tfrac{a}{t} \bmod P & t_2 &= t \bmod Q \\ t_3 &= \tfrac{a}{t} \bmod P & t_3 &= \tfrac{a}{t} \bmod Q \end{aligned}$$

But, since $\left(\frac{a}{P}\right) = \left(\frac{a}{Q}\right) = -1$, then $\left(\frac{t_1}{N}\right) = \left(\frac{t_2}{2}\right) = -\left(\frac{t}{N}\right) = -\left(\frac{t_3}{N}\right)$.

So, there is no unique $\left(\frac{t}{N}\right)$ to recover and F cannot return $\left(\frac{t}{N}\right)$ correctly more than half the time whenever a is not a square. Hence, we would have a procedure that can distinguish the two cases of $\left(\frac{a}{N}\right) = +1$, i.e. determine whether a is a square or non-square (mod N) without factoring N. This is the quadratic residuosity problem which is currently unsolved.

6.3.5.2 *Active Attack*

The protocol as described here is subject to an active attack. The attacker simply changes one encoded cryptovariable bit to a known value and replays it to a random oracle. At each attempt, the oracle will indicate correctness of the cryptovariable as a whole and the attacker can recover the key one bit at a time. There is a simple, well-known method of wrapping a protocol such as this to make it immune to active attacks (see Fujisaki and Okamoto [3]).

6.3.6 The Split Authority

Full details of the split authority protocol are described on the CESG Web site [2] and what follows is a simplistic overview. The authorities perform a one-time set-up and agree on a system modulus N (without either knowing the factorisation of N). The calculations also produce an authority secret for each party, d_1 and d_2. To generate a private key, a user contacts each authority with his identity a and the authorities will return $a^{d_1} \bmod N$ and $a^{d_2} \bmod N$ respectively. d_1 and d_2 are such that $a^{d_1 + d_2} \bmod N = b \bmod N$ where b is the user's private key. The user must simply multiply together the two 'half keys' to produce his private key.

6.4 Key Semantics

6.4.1 Traditional PKC

In traditional PKC, the public key is derived from the private key. Thus, existence of a public key implies existence of the appropriate private key. Hence, if a message is sent to a given recipient, the implication is that the recipient must be able to read that message.

It is the responsibility of the sender to ensure that the purported recipient (i.e. the recipient identified by the certificate identity) is the intended one. Verifying a certificate in the traditional model involves traversing a chain of trust, verifying each authority in turn until a party that is implicitly trusted is reached. Again, for each party, the verification keys must be obtained in order to ensure the chain of trust remains unbroken. Some of these verification keys will be obtained in the same way as the participating parties' keys, while others will be stored locally and assumed valid. Obviously, this implies that locally stored authority certificates must be protected in the same way as private keys.

In order to tie a certificate (and therefore a public key) to a user, it contains a 'distinguished name' enabling users to determine the owner of the private key. This distinguished name is often quoted as the owner's real name. This begs the question, just how many John Smiths are there and how does one differentiate between them? Furthermore, one must then associate a particular 'John Smith' with, say, an e-mail address. The fact that certificates, recipient identities and delivery agent end-point addresses are entirely disparate objects, offering little by the way of commonality, implies that traditional PKC may be prone to error. Common aliases and short-form names further complicate the matter.

6.4.2 Identifier-Based PKC

In identifier-based PKC systems, the private keys are necessarily derived from the public keys (plus some secret data). In this case, the existence of a public key is implicit and so implies nothing about the existence or validity of the associated private key. The sender is no longer able to infer anything about the validity of the recipient's 'published' public key.

The sender is not required to assert the validity of the public key since it is implicitly and strongly tied to the recipient's identity. Thus the problem of distinguishing 'John Smith' from 'John Smith' is removed if identity is based around a unique identifier (e.g. e-mail address). John.Smith@cesg.gov.uk and John.Smith@compuserve.com are obviously disparate identities. Whether or not they refer to the same person is irrelevant; the identities form identifiers of some end-point of a communication channel which are each tied to some implicit public key. Thus, if the communication end-point can be tied to a particular user, so can the public key and thus the ability to read a particular communication. This tying together of the previously disparate entities in the communication model appears to offer the possibility of a stronger trust.

6.4.2.1 Interactive Communication

Now consider the interactive communication model where the parties can exchange information bidirectionally in real time. In this scenario, no traditional signature mechanism is required since a trivial, asynchronous protocol allows for strong, mutual authentication of the end-points with no extra overhead. This protocol appears to be less vulnerable to attack since there are no concepts of 'chain of trust' or root authority certificates. The trust is placed entirely in the (aggregate) authority to issue private keys only to the intended parties and for those parties to keep safe the keys and the moduli associated with them. Compare this with the traditional PKC model, where that condition has to be met, coupled with protection of the chain of trust certificates by all parties involved, and trust of all authority parties referenced by those certificates. Security boils down to protection of a single secret in the identity PKC model.

6.4.2.2 Communities of Interest

The identifier-based PKC model allows for cryptographically separate communities of interest to be associated with the same identity since the process of obtaining a public key from an identifier involves the system modulus which can be varied to create the communities. The recipient requires a distinct private key in order to access each community of interest and each community secret (i.e. the factorisation of the system-wide modulus) is unique. Their public identity remains constant across the communities. Recall the fact that existence of a public key does not imply existence of the associated private key. Hence, if a sensitive message is accidentally sent to the incorrect recipient it cannot be read by that recipient without authorisation from the (aggregate) authority.

Knowledge of any number of private keys relevant to other communities of interest admits nothing concerning the unknown private key. This level of separation and the protection against user or routing errors appears to be impossible to achieve with traditional PKC.

6.4.2.3 *Registration*

While the practical issues associated with registration remain identical for traditional PKC and identifier-based PKC, the community of interest concept, coupled with the fact that private keys do not necessarily exist, may aid the registration process for multiple services.

Consider the following registration scenario[3]. Suppose the Government is offering a number of electronic transaction services to the citizen, that each service's transactions are conducted under its own community of interest and that the Post Office and the transacting department together form the aggregate authority for their community of interest. A user wishes to begin transacting electronically with the Government and must register to do so. He approaches the Post Office and must prove his identity to the clerk in order to obtain his token. This token may contain the Post Office 'half keys' for each community of interest and these keys will have a level of trust conferred upon them. This level of trust will be commensurate with the identity validation performed by the clerk. For this discussion let us assume that these 'half keys' have a high level of trust associated with them.

Recall that the user now needs the other 'half key' in order to construct his private key for any community of interest. Suppose that the user first wishes to transact with the vehicle licensing authority to tax his car, a service which requires little in the way of user authentication. The token already contains the Post Office 'half' of the user's private key for the appropriate community of interest. The user contacts the vehicle licensing authority and provides some basic information to identify himself. The DVLA returns his other 'half key' — without any real validation — by e-mail, for example. Obviously, this 'half key' provides little by way of strong user authentication. However, when combined with the highly trusted 'half key' on the token, the resulting private key may be trusted at some level of trust which is the combination of the two trust levels — in this case 'medium-low'.

Now consider the user transacting with the Inland Revenue and suppose that Inland Revenue transactions require a higher level of user authentication. The user may apply for his 'half key' on-line, supplying sufficient information to prove identification, and the 'half key' is sent out of band on a floppy to the registered address. This 'half key' obviously has more trust than the one supplied by the DVLA, and, when combined with the highly trusted half key on the token[4], the resulting private key has sufficient trust to be used for the Inland Revenue transactions.

6.4.2.4 *Revocation*

Revocation of an immutable, predefined public key is obviously an issue. Revocation in IDPKC systems appears to be very business-model-dependent and it seems

---

[3] It is not suggested that this is a real, sustainable registration architecture; it is merely an example.

[4] Recall that the 'half key' on the token is different for the IR and DVLA communities of interest.

that, at worst, the revocation model is equivalent to the one used with traditional PKC. In situations where one of the transacting parties is also a constituent of the aggregate authority, a locally held revocation list may be sufficient. For revocation to work in an IDPKC scenario, the concept of 'identifier' must be extended to include some variant information, e.g. the concatenation of the identifier with a one-up counter or expiry date.

If the global system rules dictate that public keys are generated from identifiers that contain the month and year (for example), then the public/private key pairs will be self-expiring at the end of each month. Similarly, if the semantics of identifier construction dictate that a sequent number is appended to the identifier, revocation at the transaction server (which is also an authority) may be sufficient for some business models.

6.5 An Example Usage Scenario

Consider an electronic pleading system for a court based around e-mail. Assume that the community for this system is limited to the court clerk, defence legal team and prosecution legal team. Each legal team consists of a barrister with a number of subordinate solicitors and paralegals.

The traditional PKC model for communication between these parties seems to work initially.

Now consider the issues when there are two John Smiths, one on the defence team and one on the prosecution team. With traditional PKC and its associated infrastructure (the PKI), there is no safety net — if a user accidentally sends prosecution information to the defence team's John Smith, the case is potentially compromised.

With IDPKC, the defence and prosecution communities would be split into cryptographic communities of interest, perhaps with the appropriate barrister and the court clerk as the authorities. There would also be a 'common' community for shared communications.

If a user in the defence community accidentally addresses a communication to the John Smith in the prosecution community, the recipient cannot read it[5] since he does not hold a private key for the defence community of interest — assuming the sender has classified his communication as 'defence only'. Thus, we add a further safety net to the communication infrastructure — two independent errors are required to leak information. This level of separation seems impossible with a traditional PKI since, in that model, existence of the public key implies existence of the associated private key.

[5] It is assumed that the court clerk and/or the defence barrister will not provide the member of the prosecution with their private key!

6.6 Summary

Identifier-based PKC is not a replacement for traditional PKC, nor is it a security panacea. However, in certain business models, an IDPKC solution may reduce cost of ownership and simplify the total user experience. It is likely that a combination PKI/IDPKC system will provide the most substantive benefits, the IDPKC portion of the system being used where the PKI will not scale and revocation is less of an issue.

Further work is required at all levels to prove the business benefits of IDPKC systems and define the methods and protocols for their interaction with traditional PKC systems.

References

1 Shamir, A.: '*Identity Based Cryptosystems and Signature Schemes*', in Blakey, G. R. and Chaum, D. (Eds): '*Advances in Cryptology*', Proceedings of CRYPTO'84, Springer Verlag (1985).

2 CESG — http://www.cesg.gov.uk/

3 Fujisaki, E. and Okamoto, T.: '*Secure Integration of Asymmetric and Symmetric Encryption Schemes*', in Krawczyk, H. (Ed): '*Advances in Cryptology',* Proceedings of CRYPTO'98, Springer Verlag (1998).

7

SECURE DIGITAL ARCHIVING OF HIGH-VALUE DATA

T Wright

7.1 Introduction

Creating, distributing and storing documents electronically has been providing greater convenience, higher speeds and increased economy for businesses and people since the advent of the personal computer. Electronic documents can be created more easily than their paper-based equivalents, transmitted around relevant parties quickly, and can be modified easily. Furthermore, electronic documents can be stored and backed up at a fraction of the cost and space of physical ones. However, one drawback has been that they have been a little too easy to copy or edit. There was no way to tell original electronic documentation from a copy, or who created it, or even whether it had been altered. This made the system rather unsuitable for auditable records, such as financial details, contracts, etc. Powerful mathematical techniques, based on cryptography, provided a solution to this, by providing protection and evidence of who created what, and whether anything had been altered. However, it is only recently that legislation has caught up with technology.

In the UK, the Electronic Communications Act 2000 enacted most of the provisions of the European Directive 1999/93/EC on a European Community framework for electronic signatures [1], so that an electronic document can be signed electronically (e.g. using a cryptographic digital signature) to bind signatories to the content they sign, with the same legal effect as if they used a traditional signature [2]. Other member states either have passed, or are in the process of passing, similar legislative measures. Wisely, the technology used is not specified; the legislation merely allows electronic techniques to have the status of paper-based standards if they can provide the same level of proof. Despite the so-called policy of technology neutrality, only public key cryptography is seriously being considered in the nascent standards to support the directive. The USA has an equivalent ESIGN bill, in addition to legislation in various states such as Utah.

Given such protection, document archives can provide enhanced protection to electronic documents. A digital archive can be defined as '... a secure repository where an authenticated owner can deposit digital data to be stored, backed up and otherwise maintained, and made accessible only to authorised users'. Any electronic data can be stored at such archives, on behalf of the originators, while cryptography can be used to keep legally binding documents here as well. This chapter describes the technologies involved in protecting documents, and how a secure digital archiving service could enhance the functionality. It then examines a possible high-level architecture for a generic service.

7.2 Technology

7.2.1 Introduction to PKIs

A public key infrastructure (PKI) is the collection of the services that are necessary to support the operation of public key cryptography. This allows participants to identify themselves and to communicate privately and free from interference across a large, open network, such as globally on the Internet, or within a corporate intranet [3].

7.2.1.1 Requirements

In order for people and businesses to feel comfortable when trading electronically using an open network such as the Internet, there are five main security features that must be provided:

- confidentiality — electronic messages that are sent must not be visible to eavesdroppers;
- authentication — communicating parties must be certain of each other's identity;
- integrity — communicating parties must know whether the data they send has been tampered with;
- non-repudiation — it must be possible to prove that a transaction has taken place;
- availability — the service must be available when required (although important, this requirement is beyond the scope of this chapter).

7.2.1.2 Keys and Algorithms

These requirements can be met by cryptographic algorithms based on mathematical keys [4].

Counter-intuitive as it may appear, security is not enhanced by using secret algorithms to encode data. There are likely to be serious flaws in most algorithms, and only by opening them to public scrutiny are these likely to be noticed.

The best way to provide security in an open, market-oriented network is to use well-known, proven algorithms, which are enabled by mathematical keys. It is the keys that must be kept secret, rather than the algorithm. Each user would then have his or her own keys, and, should they be compromised, only a new key is required. Keys may be symmetric, in which case the communicating parties would share a key.

Alternatively, they may be asymmetric, where each user has a key pair — one public and one private. The public key is made generally available, and the private key cannot be calculated from knowledge of the public key. There are many applications of these cryptographic algorithms, using either symmetric or public keys [5].

7.2.1.3 Encryption

Encryption is used to scramble messages (or any other data) so that any unauthorised parties that intercept or otherwise acquire the data cannot read them.

Data is encrypted using the public key of the recipient, so that only the recipient who owns the corresponding private key can decrypt and read the original data. Actually, for speed reasons, only a symmetric key is encrypted under the public key, and all further communication is encrypted using this shared key.

7.2.1.4 Digital Signatures

Digital signatures are used to prove who originated data, and prevent anyone surreptitiously altering it. In contrast to pen and paper signatures, a digital signature is different for each document signed, because it is mathematically bound to the data as well as to the signer.

Thus digital signatures authenticate the entity and guarantee the integrity of communications by checking if they have been altered.

For example, Bob has created a document and wishes to digitally sign it, so that everyone can tell it comes from him. He needs to create a message digest, or hash, of the document (a small representation of the whole). He encrypts this with his private key to produce a signature.

When Alice receives the signed document, she decrypts the signature using Bob's public key, to obtain the hash. She compares this to the hash she obtains by hashing the document — if equal, the signature is valid, otherwise the document was altered or the signer was not Bob, thereby checking integrity and authenticity respectively.

7.2.1.5 Role of Trusted Third Parties

The major role of a PKI is to allow entities (people, businesses, servers) to be authenticated on a network, by providing each with an identity and a means to prove it. This is provided by a service known as a certification authority (CA) [6].

A private key allows entities to perform actions unique to them, for instance, signing a document or decrypting a message. However, others must be able to match these actions to the correct identity, not just to the private key.

A digital certificate, which contains a public key and identification details, is used to match an action (on a network) to an identity. The public key is taken from the certificate and used to check the action (e.g. validating a signature). The action can then be matched to the identification contained in the certificate.

Furthermore, the trusted third party can take on a level of liability for binding public keys to the correct identity, based on the care put into the checking made before issuing certificates. This increases the general level of confidence in trading electronically.

As a first step, all the users of a PKI must register to obtain a digital identity on the network. It is the responsibility of the registration authority (RA) to perform suitable checks on the identification credentials of that entity. The level will depend on the circumstances, but any certificate should state the level of checks that have been performed. Thereafter, anyone wishing to trade with that entity can be assured of its identity and possibly also its trustworthiness.

Alice signs and sends a message to Bob. Bob obtains Alice's certificate, containing her public key and ID. He uses the public key to check that he is communicating with Alice, then uses the ID details on the certificate to determine if he wants to trust her. Thus the RA has made all the checks for Bob, and the CA lets Alice prove her identity on the network.

Of course, Bob must actually trust the CA in the first place, and be able to check the digital signatures on the certificates. The CA has its own (self-signed) certificate, which users can keep in a special store on their computers. Any signature by this CA, such as on another certificate, can then be checked and trusted, so long as this 'root certificate' is in that store. Browsers and e-mail packages come with such root certificates built into such a store, and more can be added or removed by the user. Assuming Bob trusts the CA, he can also trust the details in the certificate. However, the certificate will only be valid for a specified time (noted in the certificate) and for certain uses specified by the CA (e.g. for encryption but not for signing).

7.2.2 Revocation

Certificates are only valid for a certain time period, after which they expire — this is because hackers could try to obtain a private key from the public key in the certificate. However, a key could be compromised in other ways, such as someone else gaining access to it, or the owner could leave a company and no longer have the privileges accorded to his or her previous role. (Certificates often contain details of someone's role in addition to simple identification details.) Also, future advances in factorisation and cryptanalysis may call into question the long-term validity of a key pair that has expired at the planned end of its validity period.

There is a way of dealing with the need to prematurely retire a key called certificate revocation. When a key is compromised, the owner informs the CA, which then revokes the certificate by placing it on a list of revoked certificates, called a certificate revocation list (CRL). Now before relying on a certificate, one should first consult a CRL to see if it has been revoked. Alternatively, an on-line system could provide the current status (valid, revoked, etc) of certificates.

7.2.2.1 Key Lifetimes

Encryption and authentication keys are used, and should be managed in different ways.

In a PKI system, it is important to consider how keys and certificates are managed through their life cycles. Certificates contain information that define their validity period — after their expiry date they should no longer be trusted, and therefore signatures, etc, based on such certificates will no longer be validated. Keys and certificates should be managed according to their use (this is an important reason to separate them according to use), and is illustrated in Fig 7.1.

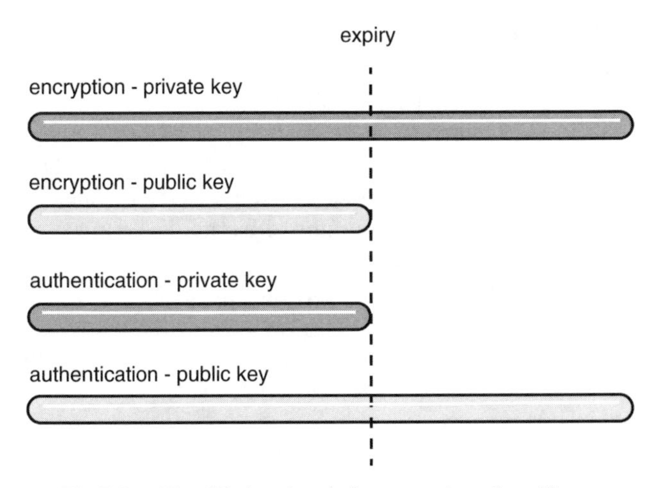

Fig 7.1 Key lifetime in relation to expiry of certificate.

In the case of encryption keys, the private keys should be kept by the owners after expiry, so that any previously encrypted messages can still be read. Hence, a personal archive for such keys is necessary. Key escrow systems operated by the CA or other Trusted Third Party can provide this. However, a public key (and the associated certificate), which is used by others to encrypt messages to the owner, should be destroyed and no longer used — and the new certificate and key made available. Using old encryption keys could compromise security. Usually software will refuse to use expired certificates for encryption.

Keys used for signing need to be handled differently. The signing private key should be immediately destroyed, as it can no longer be used to produce valid signatures and is henceforth of no further use. The signing public key, together with the certificate, needs to be kept to provide evidence for validating old signatures (typically they will be stored in the CA repository).

7.2.2.2 Timestamping

It has already been mentioned that certificates will eventually expire. This leads to a concern over keeping records intact for many years, such as financial records that are required by law to be available and auditable for many years. To keep a digital signature valid for that long, it must be provable that it was made at a date when the certificate was valid (neither expired nor revoked). This is the role of a timestamp. The format for timestamp requests and tokens is defined in standards produced by the IETF [7]. However, timestamps do not have to be based on keys. For example, Surety provides a mechanism of timestamping based on chained hashes [8].

Timestamping is a cryptographic means of proving that specific data or a document existed at a defined point in time — the document could have been created before the timestamp, but not afterwards. Thus, it would be applicable for example to patent filing (where priority in discovery is crucial), but not in proving wills (where the latest will is the legally valid one).

A further use is to enhance digital signatures by allowing a signature to be verified at a time when the signer's certificate has expired. In normal circumstances, the signature would not be verifiable after its expiry, as the private key might have been broken cryptographically or compromised before then. However, this second application emphasises one of the critical features of the service — the timestamp itself must not expire, or, at least, its lifetime must be comparable with the lifetime of the data or document it protects. This may require a renewal process.

To understand the role timestamps play in enhancing digital signatures, it is necessary to understand the procedure for verifying signatures in complex CA hierarchies as shown in Fig 7.2.

The standard verification process is as follows.

- Generate certificate chain, from signer's certificate, through intermediate CAs to root CA certificate. Each certificate is signed (by issuer) by the CA whose certificate is above in the chain, except the root CA certificate, which is self-signed.

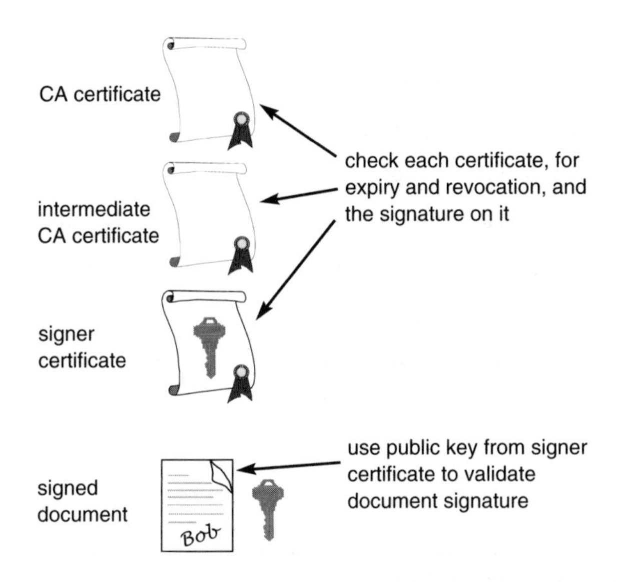

Fig 7.2 Process of validating a digital signature and its certificate chain.

- Start at root certificate (starting at signer certificate is also possible):

  — check that verifier trusts that CA explicitly (this certificate should be registered as trusted);

  — check certificate has not expired;

  — check certificate has not been revoked (not on a CRL);

  — check signature (using public key in this certificate).

- For each intermediate CA certificate and the signer certificate:

  — check certificate has not expired;

  — check certificate has not been revoked;

  — check signature using public key from previous certificate in chain.

- Use public key from signer certificate to validate document signature.

With a timestamp, the process is modified slightly. The timestamped verification process is detailed below.

- Generate certificate chain.
- Check timestamp is valid and extract time, t.
- Start at root certificate:

  — check that verifier trusts that CA explicitly;

  — check certificate had not expired at t;

— check certificate had not been revoked at *t* (not on a CRL valid at *t*);

— check signature (using public key in this certificate).

- For each intermediate CA certificate and the signer certificate:

— check certificate had not expired at *t*;

— check certificate had not been revoked at *t* (not on a CRL valid at *t*);

— check signature using public key from previous certificate in chain.

- Use public key from signer certificate to validate document signature.

It is necessary, therefore, to have a store of past CRLs against which to check whether certificates had been revoked at previous times.

7.2.3 XML

XML (extensible markup language) is a structured data description language, now commonly used for data interchange. The standards have been provided by the W3 Consortium [9] (see also Chapters 2 and 3). It has the flexibility to define digital signatures within the content, which apply just to certain sections of data; the standards for XML signatures are being defined by a W3 Consortium and IETF working group [10]. As such, XML data is not only very likely to require archiving in some of its roles, but lends itself well to storing long-term security within its structure.

7.2.4 ETERMS

The ETERMS repository is a project run by the International Chamber of Commerce to address legal issues resulting from communications in electronic commerce [11]. It allows documents to be stored and publicised, such as contractual terms incorporated into contracts 'by reference'. As the terms are stored by a trusted third party, they are not subject to later alteration. For instance, certificates could remain small but link to the ETERMS repository for the main terms of the contract. BT has participated in this project, building a trial version of the repository and looking at the integration with PKI.

7.3 Secure Digital Archiving

Secure digital archiving allows collaborative operation on documents by a number of parties, with versions stored securely at a remote site. Companies can outsource the management of documents, assured that all changes are audited by a trusted third party and access rights (read, write, etc) are strictly enforced.

Goals for all-electronic working include the replacement of all paper-based processes, keeping the ease of use and ready access provided by paper, but improving on the auditability of who accesses documents, reducing costs and improving speed of access and reliability. Digital archiving services aim to support some of these goals:

- secure storage of electronic documents, and easy retrieval by those authorised to access them;
- provision of an audit trail associated with the use of these documents (who wrote, who altered, who read, who approved — and when) in an easily accessible and legally verifiable manner.

Cryptographic techniques are important in providing a personal security environment to enable users to sign and verify their own documents, and to provide timestamping services to authenticate the existence of specific documents at a particular time. Additionally, such a service will provide:

- secure storage;
- back-ups to cover loss or alteration by virus;
- natural disasters such as fire and theft;
- strong authentication and authorisation methods to ensure that access to documents or parts of documents is restricted to specific individuals or roles.

As noted before, the timestamps must last a long time — commensurate with the time for which the documents themselves need to be stored; preferably there should be a process in place to ensure that they never expire (by, for example, reissuing timestamps).

A further requirement is that the process of verifying signatures and timestamps should be relatively simple, and, in principle, should not require access to, or reliance upon, the server. An independent verification check would be preferable in case of dispute with the service provider. Value-add services include interfaces to workflow programmes, automating approval processes, version control, and interfaces to the complete supply chain in the case of contracts for the supply of goods and services.

7.3.1 Functionality

Certain functionality needs to be provided in a digital archiving service:

- single point of access to documents;
- authentication of users, to track who is uploading or downloading documents;

- authorisation, to ensure only the desired users access restricted documents;
- encryption, to secure communications and to protect stored data;
- data protection, such as back-up and virus detection, for all data stored;
- audit trails, of who adds, reads or deletes documents;
- workflow, to allow document owners to determine how documents are accessed, stored and processed.

7.3.2 Design

The following design provides a high-level architecture that provides the necessary functionality (see section 7.3.1) for a generic service. It is based on a few major components, which are kept separate, and ideally are interchangeable, to allow a choice of vendor to provide each part.

The main components of this service are as follows:

- client software is necessary to allow users to register, log in and upload/process documents, or just view documents — a browser interface will provide most of this, but extra software will be necessary for users to sign documents to upload, or to validate signatures and timestamps on existing documents, and log-in should be over SSL, with authentication provided by client certificates or username and password;
- the digital archiving server (DAS) will authenticate clients, and provide access to timestamping services, etc (this is likely to be a Web application on a Web server supporting SSL with client authentication) — components for integrating with timestamping services, archives, authorisation databases (a list of which users can access which documents), virus checking and process flow engines will be necessary, thus possibly producing audit trails to determine who has accessed or modified documents;
- timestamping services provide the extra notarisation — either client software or the digital archiving server will access the timestamping service to obtain timestamps for digitally signed documents, use being made of any timestamping services offering the necessary compatibility;
- a data archive will actually store the data, and provide (multiple-site) back-up, disaster recovery, etc — as the data will be encrypted and protected by the digital archiving server, the archive itself just needs to store the data safely, over multiple sites, etc (PKI allows digital signatures and encryption to provide all the data protection required);
- a CRL database will be necessary to store every CRL produced, to allow validation of old signatures that have been timestamped — PKI modules on

client or DAS will access this database to check for revocation when validating timestamped signatures.

7.3.3 Processes

Figures 7.3 and 7.4 illustrate the operation of the service during uploading and retrieval, although there is some flexibility about where certain operations are carried out.

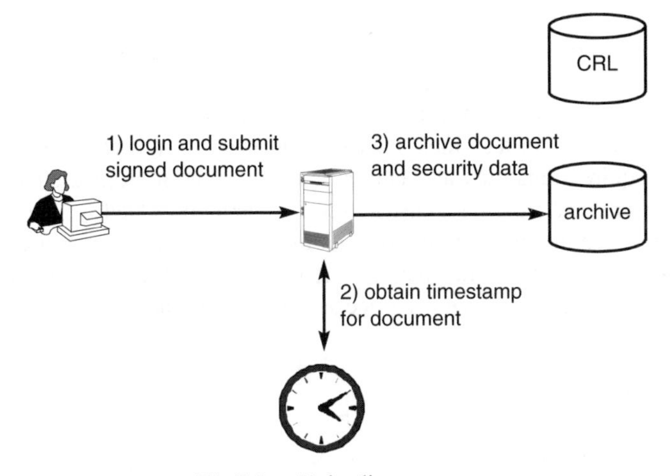

Fig 7.3 Uploading process.

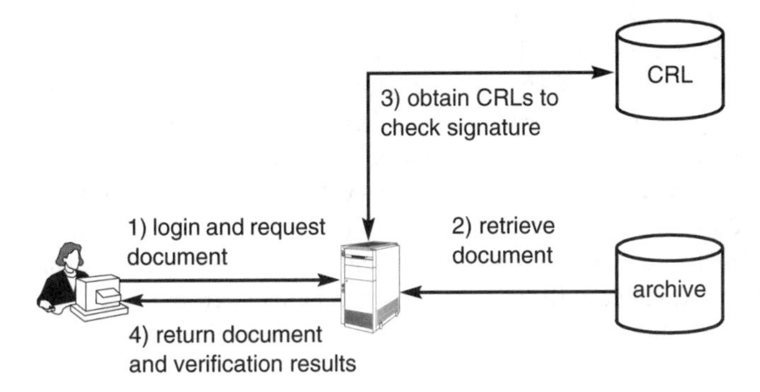

Fig 7.4 Retrieval process.

Figure 7.3 shows the uploading process. The user logs on to the service, and uploads a document, which may be signed and timestamped previously or during the upload process. It is then stored on the archive with the appropriate security data (signatures, timestamps, authorisation information).

Figure 7.4 covers access to data. A user logs in and requests a document. The service checks for authorisation and retrieves the document. The signature and timestamp can be validated (through the service or through the client), making use of the CRL store.

7.3.4 Market

This service can address the storage needs for data, ranging from financial records, where authenticity and integrity could be checked for tax, fraud or legal reasons, to contracts, deeds or intellectual property, where time and notarisation are as important as the signatories. In fact, most long-term business data, if considered important enough to require signatures and encryption, will benefit from digital archiving. There are several basic digital archiving services being provided at the moment, but as yet few possess the integration with PKI that could address the concerns of high value PKI-enhanced data. Thus there is a good market opportunity for whoever can provide a secure digital archiving service as described in this chapter.

7.4 Summary

The legal recognition of electronic signatures and other operations that can be provided by a PKI allow more important documents to exist solely electronically. The improvements in efficiency and cost are balanced by a need for secure storage and controlled access. Digital archives, integrated with PKI, can deliver a solution to this challenge, by offering secure storage for both electronic data and the accompanying security information. Further PKI services, such as timestamping, provide electronic security that will not degrade nor expire over time. Thus all the security requirements, raised by using high-value data electronically, are met. Such services are likely to be built by joint ventures between companies that can provide expertise in the relevant components, such as data storage, PKI and work-flow management. Revenue will be driven by the demand from a growing market for handling high-value data that has recently moved to solely electronic format.

References

1 Official Journal of the European Communities, pp L13-12 to L13-20 (January 2000).

2 Electronic Communications Act 2000 — http://www.legislation.hmso.gov.uk/acts/acts2000/20000007.htm

3 Ford, W. and Baum, M.: '*Secure Electronic Commerce*', (2nd ed) Prentice Hall (2001).

4 Phoenix, S. J. D.: '*Cryptography, trusted third parties and escrow*', in Sim, S. and Davies, J. (Eds): '*The Internet and Beyond*', Chapman & Hall, London, pp 62-95 (1998).

5 Schneier, B.: '*Applied Cryptography*', (2nd ed) John Wiley and Sons, New York (1995).

6 Skevington, P. J. and Hart, T. J.: '*Trusted third parties in electronic commerce*', in Sim, S. and Davies, J. (Eds): '*The Internet and Beyond*', Chapman & Hall, London, pp 51-61 (1998).

7 PKIx Draft — Internet PKI Part V: Time Stamp Protocols — http://search.ietf.org/internet-drafts/draft-ietf-pkix-time-stamp-09.txt

8 Surety Technologies Inc — http://www.surety.com/

9 W3C Standard for XML — http://www.w3.org/XML/

10 W3C XML Signature Working Group — http://www.w3.org/Signature/

11 Mitrakas, A.: 'The proposed ETERMS Repository of the International Chamber of Commerce', EDI Law Review, **3**(4) (1997) — http://kapis.www.wkap.nl/oasis.htm/135070

8

WIRELESS SECURITY

C W Blanchard

8.1 Introduction

User expectations for instant communication and ease of use, over terminals which are easily lost or stolen, present a number of unique challenges in implementing security in the mobile environment. As with most other systems, the main objective of security is to maintain integrity, privacy and confidentiality and to prevent fraud and denial of service. Examples of these challenges are:

- access and use of the service to avoid or reduce a legitimate charge;
- loss of confidentiality and/or integrity of user and operator data;
- denial of a user's access to their service;
- denial of access by all users to a service.

GSM was designed from the beginning with security in mind and has stood up well to the type of attacks perceived to be likely at the time due to the fact that:

- the responsibility for security is with the home environment (HE) operator;
- the HE operator can control the use of the system by the provision of the subscriber identity module (SIM) which contains a user identity and authentication key;
- the long-life authentication key is not required by the serving network (SN) when roaming, hence this key is neither exposed over the air nor exposed across the interface between the SIM and the mobile;
- the level of trust the HE operator needs to place in the user, serving network and manufacturer of the mobile equipment (ME) is kept to the minimum.

The development of a new security architecture for 3G [1, 2] was based on an evolution of the GSM security architecture and includes the following requirements:

- additional features needed for actual or predicted change in the operating environment;

- additional or enhanced features to overcome actual or perceived weaknesses in 2G;
- maintain compatibility with GSM wherever possible and retain those features of GSM that have proved to be robust and useful to the user and network operator.

The following enhancements to the GSM security model were seen as a priority for 3G:

- mutual authentication;
- assurance — to provide assurance that authentication information and keys are not being reused (key freshness);
- integrity — to ensure the protection of specific signalling messages, for example to secure the encryption algorithm negotiation process;
- encryption — use of stronger encryption (a combination of key length and algorithm design), and termination of the encryption further into the core network to encompass microwave links.

8.2 Security Mechanisms in 3G for the CS and PS Domain

8.2.1 SIM-Based Authentication

One of the most important security features is that of the subscriber identity module (SIM). The SIM is a removable security module inserted into a terminal device and is issued and managed by the HE operator. As the security is wholly contained within the SIM it ensures a security model that is independent of the terminal. Therefore, if a terminal is stolen and the SIM is removed and inserted in another terminal, the HE operator has the power to prevent fraudulent calls based on the information provided by the SIM and does not need to have any information about the terminal device. This concept has been the most significant in maintaining the security of GSM, while retaining general user acceptance of the service.

The 3G system retains the challenge and response authentication mechanism used in 2G. This mechanism is based on a symmetric secret authentication key shared between the SIM and the authentication centre (AuC) in the home environment.

The serving network does not require the authentication key and algorithm used to calculate responses to a challenge. This helps to keep the level of trust placed in the (many possible) serving networks to a minimum.

This method also allows the challenge-response algorithm to be made specific to the home environment, which means that the impact of any compromise can be confined to the user base of just one operator rather than the entire user community world wide. However, the GSM authentication mechanism is only unidirectional, therefore the user is not given the assurance that they have established a connection with an authentic serving network.

Within 3G, consideration was given to making use of the work done in the IP world, e.g. public-key asymmetric techniques. However, the infrastructure was already in place within GSM to allow global roaming and recognition of security information. This is not the case with public key systems and a truly global public key infrastructure (PKI) has many inherent problems to address. In addition, concerns were aired regarding whether the SIM card would be able to handle an asymmetric key protocol and still retain the performance to which the users have become accustomed.

Within 3GPP, a technique known as the authentication and key agreement (3GPP AKA) mechanism was conceptualised and designed, as shown in Fig 8.1. This mechanism provides two enhancements to ensure the provisioning of integrity and mutual authentication.

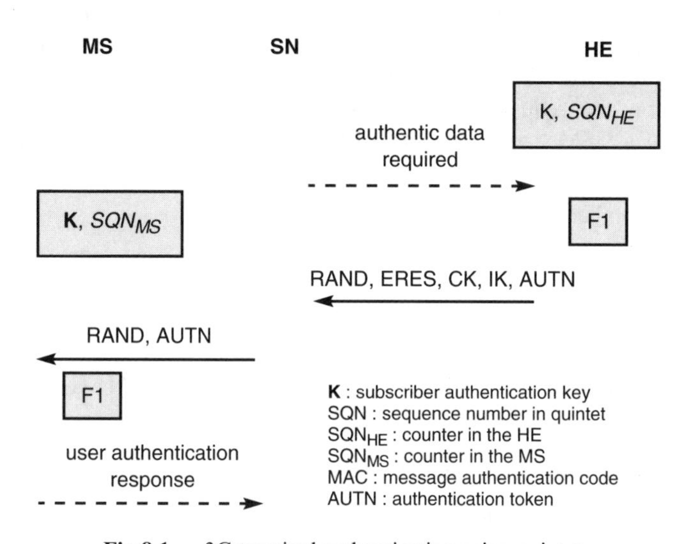

Fig 8.1 3G terminal authentication using quintets.

The integrity protection of signalling messages and the ciphering algorithm are described later.

Additional parameters and cryptographic checks are introduced to provide mutual entity authentication and the establishment of a shared secret cipher key and integrity key between the universal mobile telephony system SIM (USIM) at the user side and the home location register/authentication centre (HLR/AuC) at the network side. The AKA mechanism uses symmetric key techniques using a secret subscriber authentication key **K** that is shared between, and available only to, the USIM and the HLR/AuC in the user's home environment. In addition, the AuC keeps track of a counter SQN_{HE} and the USIM keeps track of a counter SQN_{MS} and stores additional data to support network authentication and to provide the user with assurance of key freshness.

The mutual authentication mechanism is based on a challenge/response protocol identical to the GSM subscriber authentication and key establishment protocol, combined with a sequence number-based one-pass protocol for network authentication. This protocol was derived from the ISO standard ISO/IEC 9798-4.

8.2.2 Distributing Authentication Data to Support Global Roaming

As in GSM, a user wishing to make or receive calls on their handset while travelling abroad must be authenticated by the home environment via the serving network. This is achieved in 2G by the transfer of sets of authentication data from the HE to the SN. For 3G, this mechanism is enhanced to give greater control over when and how this data is used by the serving network. The procedure for distribution of authentication data from the HE to SN starts with the visited location register (VLR) or serving gateway serving node (SGSN) sending a request to the user's HLR/AuC. Upon receipt of that request the HLR/AuC sends an ordered array of *n* quintets (the equivalent of a GSM 'triplet') to the VLR or SGSN. Each quintet consists of the following components:

- a random number challenge — RAND;
- an expected response — XRES;
- a cipher key — CK;
- an integrity key — IK;
- an authentication token — AUTN = [SQN ⊕ AK] || AMF || MAC-A.

To create these quintets the HLR/AuC:

- generates a fresh sequence number SQN from a counter SQN_{HE};
- generates an unpredictable challenge RAND;
- computes a message authentication code for authentication MAC-A = $f1_K$ (SQN || RAND || AMF) where f1 is a message authentication function;
- computes an expected response XRES = $f2_K$ (RAND) where f2 is a (possibly truncated) message authentication function;
- computes a cipher key CK = $f3_K$ (RAND) where f3 is a key generating function;
- computes an integrity key IK = $f4_K$ (RAND) where f4 is a key generating function;
- computes the anonymity key AK = $f5_K$ (RAND) where f5 is a key generating function (AK is used to conceal the sequence number as the latter may expose the identity and location of the user — the concealment of the sequence number is to

protect against passive attacks only, and therefore if no concealment is needed then f5 ≡ 0 (AK = 0));

- assembles the authentication token AUTN = SQN [⊕AK] || AMF || MAC-A and the quintet Q = (RAND, XRES, CK, IK, AUTN) and updates the counter SQN_{HE}.

Each quintet is good for one authentication and key agreement between the VLR or SGSN and the ME/USIM:

- when the VLR or SGSN initiates the over-the-air authentication and key agreement procedure, it selects the next quintet from an array held in the VLR and sends the parameters RAND and AUTN to the user;
- the USIM checks whether AUTN can be accepted and, if so, produces a response RES that is sent back to the VLR or SGSN — the USIM also computes a session cipher key (CK) and an integrity key (IK);
- the VLR or SGSN compares the received RES with XRES, and, if they match, the VLR or SGSN considers the authentication and key agreement exchange to be successfully completed and selects the corresponding CK and IK from the quintet;
- the established keys CK and IK will then be transferred by the USIM and the VLR or SGSN to the entities which perform ciphering and integrity functions, i.e. the ME at the user side and the RNC at the network side.

8.2.3 Authentication and Key Agreement in the USIM

The processing in the USIM upon receipt of a (RAND, AUTN) (Fig 8.2) is as follows.

- if the sequence number is concealed, the USIM computes the anonymity key AK = $f5_K$(RAND) and retrieves from AUTN the unconcealed sequence number SQN = (SQN ⊕ AK) xor AK;
- the USIM then computes XMAC-A = $f1_K$ (SQN || RAND || AMF) and compares XMAC-A with MAC-A included in AUTN;
- if they are different, the USIM triggers the ME to send back a 'user authentication response' with indication of integrity failure to the VLR or SGSN and abandons the procedure — the next stages are for the case where XMAC-A and MAC-A are equal;
- next the USIM verifies that the received sequence number SQN is acceptable (the HE has some flexibility in the management of sequence numbers, but the verification mechanism needs to protect against wrap around and allow, to a

certain extent, the out-of-order use of quintets) — there is a detailed description of a mechanism to generate sequence numbers that satisfy all conditions [1];

- if the sequence number SQN is not acceptable, the USIM computes the re-synchronisation token AUTS and triggers the ME to send back a 'user authentication response' to the VLR or SGSN, with an indication of synchronisation failure, including the resynchronisation token AUTS, and abandons the procedure — the remaining paragraphs therefore apply for the case where SQN is acceptable [1];
- the USIM then computes the response RES = $f2_K$(RAND) and triggers the ME to send back a user authentication response back to the VLR or SGSN, with an indication of successful receipt of the signed challenge and including the response RES;
- finally, the user computes the cipher key CK = $f3_K$ (RAND) and the integrity key IK = $f4_K$ (RAND).

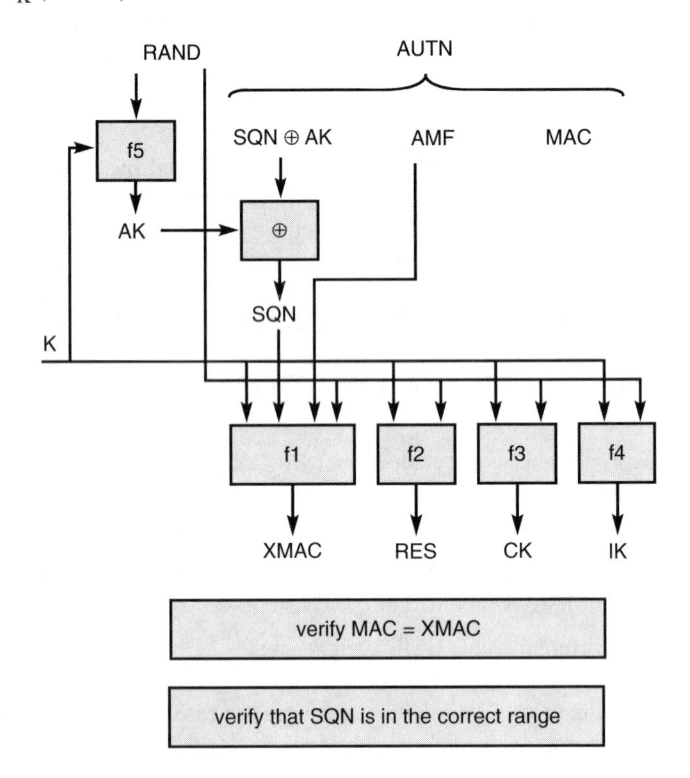

Fig 8.2 Authentication and key agreement in the USIM.

The USIM keeps track of an ordered list of the highest batch number values it has accepted. Using this list mechanism, it is not required that a previously visited SN/

VLR deletes the unused authentication vectors when a user deregisters. Retaining the authentication vectors for use when the user returns later may be more efficient as regards signalling. The USIM accepts the sequence number if it is not already on this list or if it is greater than the highest value in the list. In practice, the decision is more complex, e.g. forced wrap-around of the counter must be prevented, by ensuring that the USIM will not accept arbitrary jumps in batch numbers. If the sequence number received in an authentication request is accepted, the list is updated. If a sequence number received in a user authentication request is rejected, the list remains unaltered.

The integrity key IK and the cipher key CK are then used to protect subsequent signalling and user information between the radio handset and radio network controller (RNC) in the core network.

8.2.4 Confidentiality of User Traffic on the Air Interface

Air interface encryption is being retained but not just for the benefit of the user (Fig 8.3). This confidentiality is actually essential for the network operator to be able to ensure that the validity of the authentication at the start of the call is maintained throughout the call, i.e. to prevent a session from being hijacked. It proved impossible to reach an agreement to make this a mandatory feature, due to concern about restrictions on the use of encryption in some countries. The use of integrity protection on the signalling messages can be an alternative means of achieving this end.

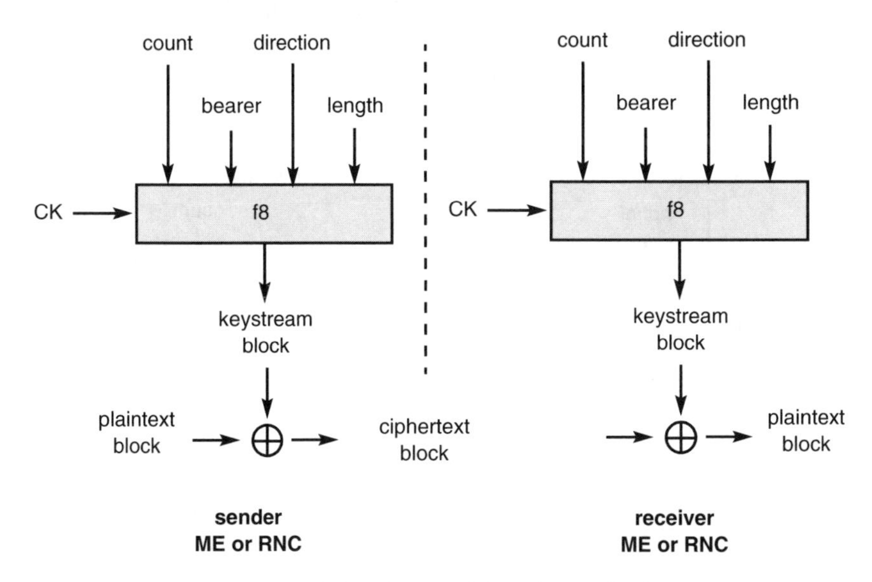

Fig 8.3 Air interface confidentiality mechanism.

Protection of user data is by means of an encryption algorithm, function f8, and used for the protection of user and signalling data sent over the radio access link between RNC and ME. f8 is a symmetric synchronous stream cipher. f8 is used to encrypt plaintext by applying a keystream using a bitwise XOR operation. The plaintext may be recovered by generating the same keystream using the same input parameters and applying it to the ciphertext using a bitwise XOR operation. For 3G Release 99, f8 is based on the Kasumi algorithm [3].

8.2.5 Integrity Protection of Signalling Messages

The approach to the negotiation and initialisation of ciphering has been carefully considered with integrity protection applied to the final security mode command that starts the ciphering. Any intruder attempting to take advantage of users in networks where ciphering is not applied, would have to overcome the mandatory integrity protection to hijack the connection. In networks where ciphering is applied, the intruder might attempt to spoof a message turning it off, but would be prevented by the integrity protection mechanism.

The receiving entity (MS or SN) must be able to verify that signalling data has not been modified in an unauthorised way since it was sent by the sending entity (SN or MS). Also, it must be ensured that the data origin of the signalling data received is indeed the one claimed. This is achieved by the inclusion of a data integrity function for signalling data. GSM does not have this functionality. The message authentication code (MAC) function f9 is used to authenticate the data integrity and data origin of signalling data transmitted between the mobile equipment (ME) and the radio network controller (see Fig 8.4). The MAC function f9 is allocated to the ME and the RNC. For 3G Release 99, f9 is based on the Kasumi algorithm [3].

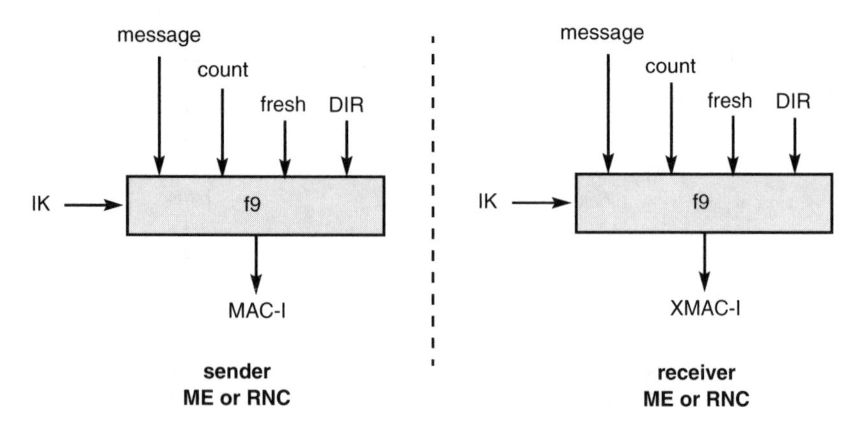

Fig 8.4 Air interface integrity mechanism.

8.3 Security Mechanisms in 3G for the IM Domain

8.3.1 Authentication by Shared Secret

The IP multimedia core network (IM CN) subsystem enables mobile network operators to offer their subscribers multimedia services based on and built upon Internet applications, services and protocols. It uses the IETF-defined session initiation protocol (SIP) as the signalling protocol between the network and the user equipment and between elements within the network [4]. An overview of the functional architecture is given in Fig 8.5.

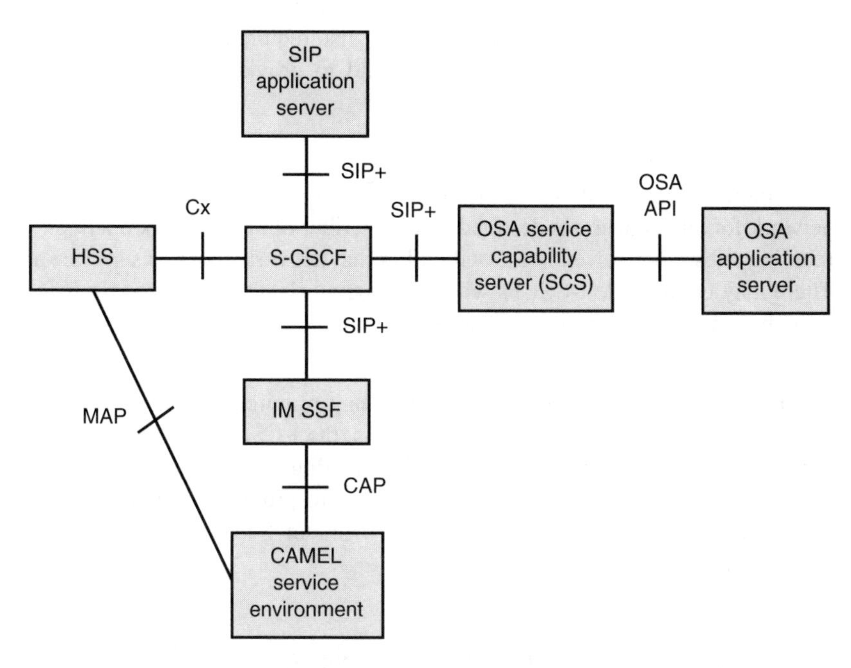

Fig 8.5 Functional architecture for the provision of service in the IMS.

In the PS domain, service is not provided until a security association is established between the mobile equipment and the network.

The IM CN subsystem is an overlay to the PS domain and may be operated by a party that is independent of the operator of the PS domain. Hence, a second security association is required between the multimedia client (UA or UE) and IM CN subsystem before access is granted to multimedia services. However, even in this case it may be that there is a trust relationship between these two independent operators and this second layer of authentication, although provided by the supplier in a standardised way, may not be invoked for some subscribers.

A similar argument can be made for WAP security, in that a second authentication to a WAP gateway owned and managed by the network operator is unnecessary.

The call state control function (CSCF) is actually split into three component parts, each with a defined role.

- Proxy-CSCF

  The proxy-CSCF (P-CSCF) is the first contact point within the IM CN subsystem. Its address is discovered by UEs following PDP context activation in the PS domain. The P-CSCF accepts requests and services them internally or forwards them on, possibly after translation. The P-CSCF may also behave as a user agent — for example, in abnormal conditions it may terminate and independently generate SIP transactions. It is also required to maintain a security association between itself and each UE, and to provide security towards the serving-CSCF.

- Interrogating-CSCF

  The interrogating-CSCF (I-CSCF) is the contact point within an operator's network for all connections destined to a subscriber of that network operator, or a roaming subscriber currently located within that network operator's service area. There may be multiple I-CSCFs within an operator's network. Its main function is to hide the configuration, capacity, and topology of the network from the outside.

  When the I-CSCF is chosen to meet the hiding requirement, then, for sessions traversing across different operators' domains, the I-CSCF may forward the SIP request or response to another I-CSCF allowing the operators to maintain configuration independence. It is also required to provide security towards the proxy-CSCF, as defined by the Network Domain Security Specification TS 33.200 [5]

- Serving-CSCF

  The serving-CSCF (S-CSCF) performs the session control services for the UE. It maintains a session state as needed by the network operator for support of the services. Within an operator's network, different S-CSCFs may have different functionality.

- Reusing 3GPP AKA in the IP multimedia world

  The P-CSCF terminates IMS access confidentiality and integrity protection of SIP messages from the UE. Security associations needed to achieve this, e.g. CK and IK, are user specific here and are established by the 3G AKA mechanism, which is being reused for IM-services [6]. This will mean that, in the future, the USIM AKA application may be deployed in more than just cellular handsets, and will extend to a large range of multimedia clients.

Security associations between P-CSCF and S-CSCF are established via network domain security mechanisms [6] and are not user specific.

It is still undecided whether authentication should terminate in the HSS in the home network or the S-CSCF. The advantages of terminating the authentication in the S-CSCF are currently being considered in 3GPP [7]. The current working assumption is that authentication will be terminated in the S-CSCF:

- in GSM and 3G CS and PS, the HSS is stateless and, if the AKA is terminated in the S-CSCF, this stateless concept could be maintained (the HSS would just be a (transaction-) stateless server, which responds to queries), whereas if the AKA were terminated in the HSS, the HSS would have to send out requests and wait for responses, for a potentially large number of users simultaneously — this could reduce HSS performance significantly;
- the HSS may become more vulnerable to denial of service (DoS) attacks due to the need to keep the state of the actual running SIP registration transactions of a possibly very large number of users;
- it would be possible to pre-compute more than one authentication vector (quintet) and to send a batch of pre-computed quintets to the S-CSCF — the S-CSCF could then handle SIP transactions, which require re-authentication autonomously, without the need to contact the HSS each time.

However, the following disadvantages have also been identified:

- the home network would become more vulnerable to DoS attacks as the S-CSCF is selected before the UA is authenticated;
- the S-CSCF temporarily needs to store authentication related information (this is not the case, when authentication is carried out in the HSS) — the S-CSCF has to keep user state anyhow, so may not have much impact on performance and capacity. It is expected that there will be a few HSSs handling a very large number of users, but many more S-CSCFs.

An example of the information flow for authentication terminating in the S-CSCF is given in Fig 8.6.

The security-related information flow is as follows (see 3G TS 23.228 for details of steps 1 to 8 [4]):

9 the S-CSCF sends a *Cx-Put* to the HSS;

10 the HSS stores the association between subscriber identity and S-CSCF address and sends back a *Cx-Put Resp*;

11 the S-CSCF sends a request for authentication data *Cx-AuthDataReq* to the HSS;

12 the HSS selects a quintuple with user specific authentication data RAND||AUTN||XRES||CK||IK;

13 in an *Cx-AuthDataResp* message the HSS sends the quintuple RAND||AUTN||XRES||CK||IK to the S-CSCF;

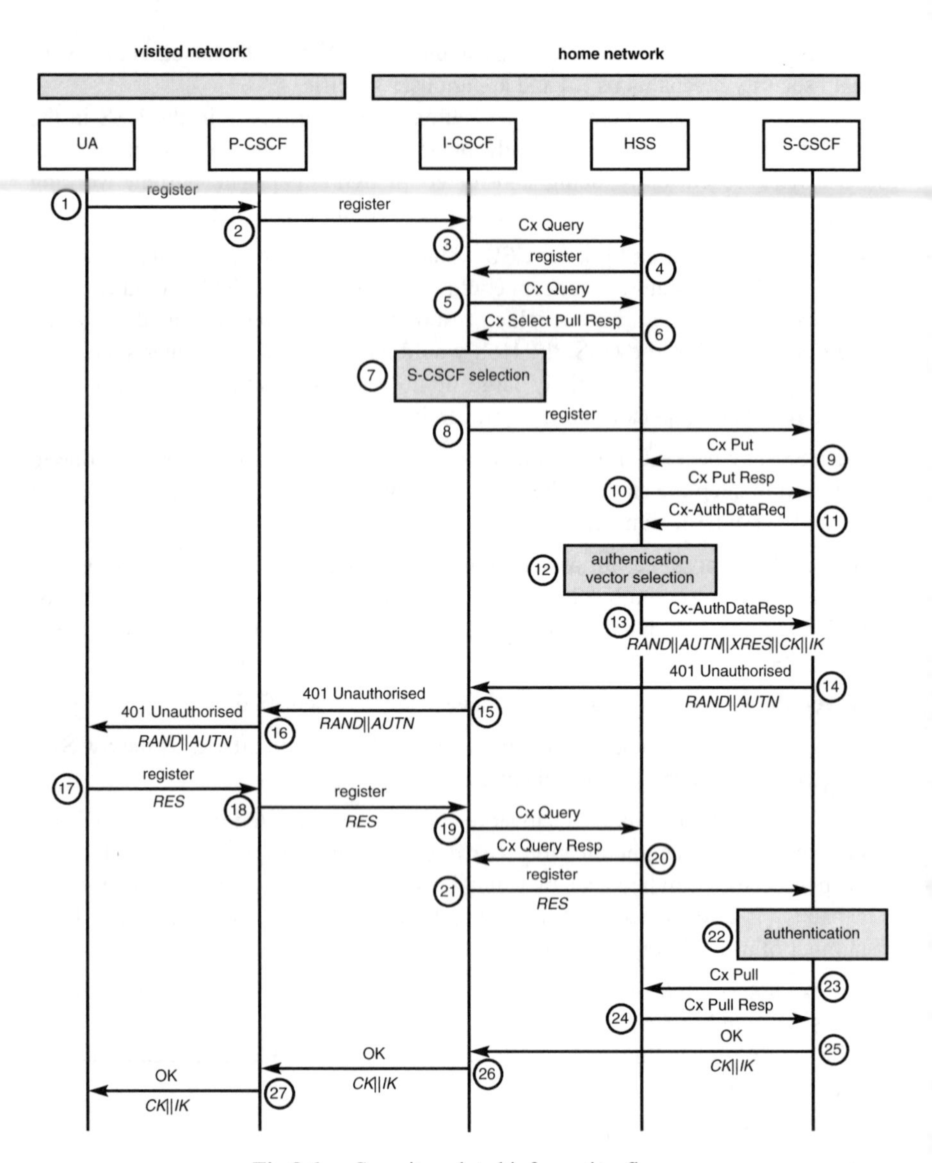

Fig 8.6 Security-related information flow.

14 the S-CSCF sends a *401 Unauthorised* message to the I-CSCF in order to indicate that the registration requested by the UA needs to be authenticated — this message contains the parameters RAND and AUTN which are needed for authentication purposes in the UA;

15 the I-CSCF forwards the received message (including the parameters RAND and AUTN) to the P-CSCF;

16 the P-CSCF forwards the received message (including the parameters RAND and AUTN) to the UA;

17 the UA checks AUTN, computes the authentication response RES and sends RES in a register message to the P-CSCF;

18 the P-CSCF forwards the received message (including the parameter RES) to the I-CSCF;

19 the I-CSCF sends a *Cx-Query* to the HSS;

20 the HSS sends a *Cx-Query Resp* to the I-CSCF with the address of the S-CSCF;

21 the I-CSCF forwards the received REGISTER message (including the parameter RES) to the S-CSCF;

22 the S-CSCF authenticates the user by checking if the received value RES and the stored value XRES are equal — if yes, then the UA is successfully authenticated;

23 S-CSCF sends a *Cx-Pull* to the HSS;

24 the HSS sends a *Cx-Pull Resp* to the S-CSCF;

25 the S-CSCF indicates to the I-CSCF that authentication was successfully completed by sending an OK message, which includes the session keys IK and CK for integrity/confidentiality protection of SIP signalling;

26 the I-CSCF forwards the received message (including the parameters CK||IK) to the P-CSCF;

27 the P-CSCF sends an *OK* to the UA (which does not include the parameters CK||IK).

8.4 Is There a Role for PKI in 3G?

Authentication and key agreement for the CS, PS and the current proposal for the IM domain, all use shared secret symmetric keys, rather than public/private key pairs. Replacement of this current global infrastructure, which is critical in order to support roaming in GSM, with a PKI system that provides the same reach, performance and security, would take many years.

Performance is a critical issue in a cellular radio system, where the overhead for protecting network signalling during handover between cells in say, a rapidly moving vehicle, must be kept to the absolute minimum. Comments, from supporters of public key, that shared secret symmetric key systems do not scale, are often not accepted as particularly relevant to the argument, with the authentication systems described in this chapter supporting 43 million users in the UK and some 1 billion world-wide by 2003.

Some specialist applications and services running over the top of 3G mobile networks may, however, require digital signatures and non-repudiation services, where there is no alternative to public key. These services are particularly relevant for MExE devices where the security framework relies on a signature being appended to certain downloaded applications which in turn need to be verified and validated — the so-called 'signed-content' model.

8.4.1 In an Ideal World

So how can these services be supported across the global GSM/3G infrastructure? A recent contribution to 3GPP suggests a possible solution [8]. Firstly, the contribution describes the ideal global PKI model, where the USIM would be enabled with a signing key pair at personalisation time. The CA certifies the public key component of this key pair over the air and the resulting public key certificate is placed in a directory service application. It is also stored in a key back-up server. 'Service certificates' would only be issued to provide access control to a service, with restricting parameters defining type of service and length of service. The service request would be signed by the UE's private signing key. The 'service certificate' would be generated and signed by the network CA.

For future signature verification purposes, only the public key certificate of the UE's signing key would need to be stored. Since this key would have a relatively long life, it would considerably reduce the key storage requirements.

8.4.2 A Possible Compromise

The contribution [8] then suggests that the major stumbling block to the introduction of these specialist services running over the top of 3G will be the lack of a large-scale infrastructure to authorise, and charge, users for such advanced services. Such an infrastructure could be built from scratch, but this could take years. Alternately, by making some minor additions to the 3GPP-security architecture, it is likely that charging for these services could be bootstrapped over the existing authentication and billing infrastructure in place for GSM and 3G.

As a first step towards this goal, it is proposed to 'bootstrap' a local public key infrastructure (PKI) from the cellular infrastructure. This is of course very different from traditional approaches to using public key technology, which presumes a global PKI and trust infrastructure. The justification for the new approach is that attempts to build such a global infrastructure have so far not succeeded and the proposal for 'bootstrapping' a local PKI has a better chance of success.

It uses the 3GPP authentication and key agreement process that results in an integrity key (IK) between the user equipment (UE) and the serving network to create an authenticated channel to submit the UE's public signature verification key and obtain a temporary certificate issued by the serving network. There is, of course, a security window of vulnerability for this public key, prior to certification, since integrity protection would normally need to be provided to protect the public key from substitution. The public key submission or certificate request would be signed by the UE's private signing key to ensure that the correct private key is being certified.

It is likely that this would then be used for subscription services for single or multiple purchases during the validity period. The information about the type of use allowed can be put into the signing procedure. It can also be used for subscription-based access control with charges resulting from the service which can be added to the user's mobile telephone bill.

It is proposed to add a pair of new signalling message types for 'certificate request/response'. The serving network element should recognise the certificate request message and route them towards the local certification authority (CA). Similarly it should recognise the certificate responses and route them towards the UE. In addition, to support long-term public keys, it is proposed that there be an extra 160-bit field in the authentication data response of the 3GPP authentication and key agreement protocol, intended to convey a public key digest from HE/HLR to the visited network.

8.5 Summary

This chapter has described the security architecture for 3G that maintains compatibility with GSM as far as possible. Those features of GSM that have proved to be robust and useful have been retained, but enhanced to overcome actual or perceived weaknesses in 2G systems. These features are now being designed into 3G equipment that will be deployed across the world in the next two years. The chapter then described why these features are based on shared secret key techniques and how these techniques may extend into the wired multimedia world, until such time that a truly global PKI is a reality. Chapter 9 presents an alternative view based on the work of the WAP Forum.

References

1 Technical Specification 3rd Generation Partnership Project; Technical Specification Group Services and System Aspects; 3G Security; Security Architecture (3G TS 33.102 version 3.7.0 Release 1999) — http://www.3gpp.org/

2 USECA Homepage — http://www.useca.freeserve.co.uk/

3 3GPP Confidentiality and Integrity Algorithms F8 and F9 — http://www.etsi.org/dvbandca/3gpp/3gppspecs.html

4 3GPP TSG SA WG2 Architecture, TS 23.228: '*IP Multimedia (IM) Subsystem — Stage 2*', v. 2.0.0 (March 2001).

5 3GPP TSG SA WG3 TS 33.200: '*Network Domain Security (Release 4)*', v0.3.2 (2001-02) S3#17, Gothenburg, Sweden (S3-010055) (March 2001).

6 3GPP TSG SA WG3 Security, TS 33.203: '*Access security for IP-based services (Release 5)*', v 0.2.0 (March 2001).

7 3GPP TSG SA WG3: '*Alternatives for terminating authentication in the home domain of the IM Subsystem*', Siemens AG, Madrid (S3-210053) (April 2001).

8 3GPP TSG SA WG3: '*Support of certificates in 3GPP security architecture*', Nokia with comments by Orange, S3#17, Gothenburg, Sweden (S3-010040) (March 2001).

9

ADAPTING PUBLIC KEY INFRASTRUCTURES TO THE MOBILE ENVIRONMENT

N T Trask and S A Jaweed

9.1 Introduction

A prerequisite for many commercially attractive services is the deployment of a secure infrastructure protecting the interests of all the parties. For many of the mobile operators and service providers the PKI model is of particular interest, offering the authentication, integrity, non-repudiation and confidentiality requirements demanded by many application providers. Financial institutions, for example, need to offer their customers the ability to bank on-line or trade in shares using their mobile devices, while retaining the same level of transaction security available in the 'wired world'.

Currently under GSM, there are two common approaches to providing value-added services on top of bearer services — via SIM Toolkit (STK) or via the use of wireless application protocol (WAP). There are already a number of STK-based PKI solutions available from vendors today, but while they may offer early-to-market advantage, they are typically proprietary, 'walled-garden' solutions. The remainder of this chapter focuses on WAP, and discusses the WAP Forum's standards-based approach to implementing a wireless PKI — an initiative that is proving attractive to many operators.

9.2 WAP Overview

The wireless application protocol (WAP) is a *de facto* industry standard defined by participants in the WAP Forum [1]. It allows Internet content to be delivered to wireless devices such as digital mobile telephones and personal digital assistants (PDAs). In the 'wired world', Web servers deliver data in hypertext markup language (HTML) format to Web browsers. In contrast, in the 'wireless world', wireless markup language (WML) content is served to wireless devices via WAP gateways.

Wireless networks, and the mobile clients available today, impose many constraints not normally experienced in the wired world. Typically these include narrow wireless bandwidth, high message latency, limited client computational capabilities, low memory resources, low processing power and limited user interfaces. It is expected that, in time, these constraints will diminish. Currently, WAP, and specifically the WAP PKI, includes features which address this restricted working environment, while leveraging existing standards.

It is useful when discussing the security features available in the WAP environment to position these against the WAP protocol stack. Figure 9.1 depicts this stack, as defined in the specification [2]. It is a layered architecture, where all of the layers can be accessed by the layers above, by means of well-defined interfaces, as well as by other services and applications.

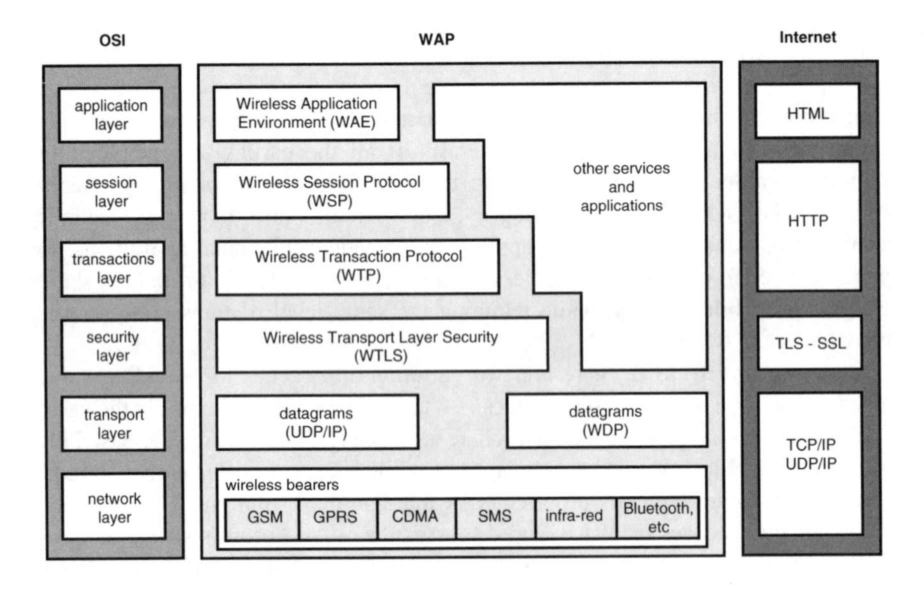

Fig 9.1 The WAP protocol slack [1].

It is important to note that the WAP protocols were designed to operate over a variety of different bearer services — not just GSM, but also GPRS and beyond.

Figure 9.1 positions the layers of the WAP stack against a pseudo-ISO OSI stack. It also compares WAP protocols with parallel protocols in the wired Internet environment. This composite diagram shows that, as in the wired case, security services are available at different layers in the WAP stack. At the application layer, WAP supports digital signatures (using the WMLScript 'signText' function), while at a lower layer it defines the wireless transport layer security (WTLS) protocol — analogous to transport layer security (TLS) or secure sockets layer (SSL) [3].

9.3 WAP Security

WTLS [4] is the wireless security protocol which enables client and server authentication through the use of digital certificates. There are three levels of security provision at various stages of adoption. WTLS Class 1 provides confidentiality and data integrity between the wireless device and the WAP gateway. Class 2 adds authentication of the WAP gateway to the security services provided by Class 1. Finally, Class 3 builds on Class 2 by adding support for authentication of the wireless client.

The scripting language developed for the WAP environment, WMLScript, includes a function that supports the digital signature of WML coded content [5]. This 'signText' function allows a wireless user to digitally sign a transaction in a way that can be verified by a content server. This provides end-to-end authentication of the client, together with integrity and non-repudiation of the transaction.

In order to leverage these security features, it is clear that a WAP public key infrastructure (WPKI) is required [6], implemented in a similar way to the more established Internet PKIs.

9.3.1 The WAP Security Architecture

Figure 9.2 illustrates the WAP security architecture. It should be noted that wireless clients are connected to services in the wired Internet via a WAP gateway. The WAP gateway is an access point to the fixed Internet for wireless devices, converting both data and protocols from the wireless to the wired network. At this point, at the transport layer, there is a security protocol translation between WTLS and SSL.

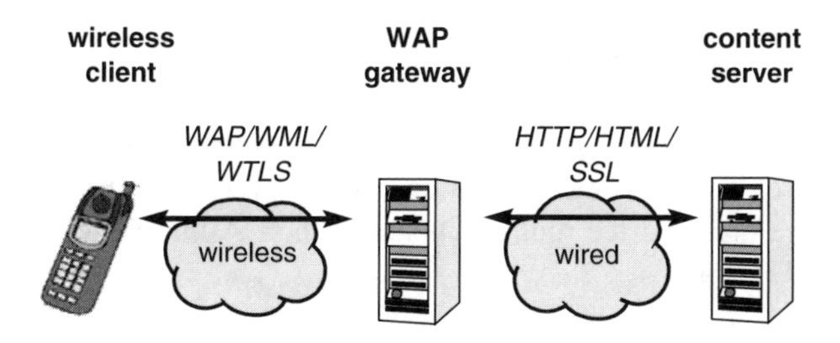

Fig 9.2 The WAP security architecture, showing how the WAP gateway acts as an intermediary between the wireless client and content server.

9.3.2 Authentication of the Content Server

Some applications may require data to be exchanged securely between the wireless client and content server. Because of the conversion of protocols at the WAP gateway, it is not possible at this time to provide true end-to-end encryption. Instead, the link has to be secured in two parts — WTLS is used to protect traffic between the wireless client and the WAP gateway, while in the wired Internet conventional SSL is used between the gateway and content server. Until the adoption of a solution such as described in the WAP Forum document [7], which proposes transport layer end-to-end security, there is a clear vulnerability at the WAP gateway as encrypted data is deciphered to clear text before being re-enciphered using SSL.

In establishing the SSL connection, the content server authenticates itself to the WAP gateway by passing over its digital certificate. This gateway must therefore hold the root certificate of the CA that issued the server's certificate (see Fig 9.3).

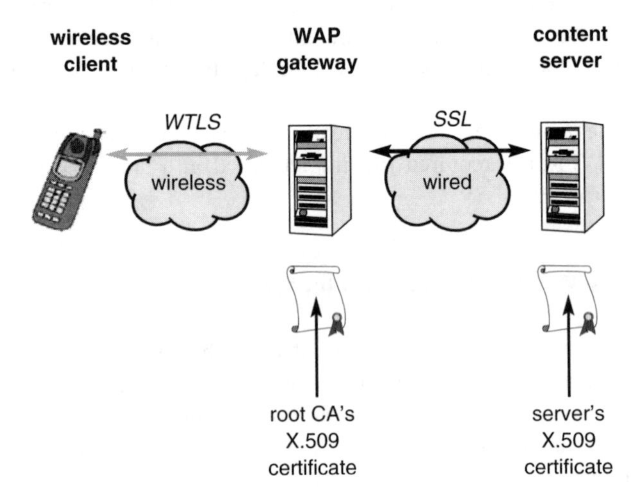

Fig 9.3 Securing the Internet connection using SSL.

9.3.3 Authentication of the WAP Gateway

In a very similar way, when establishing the WTLS connection, the WAP gateway authenticates itself to the wireless client by passing over its digital certificate. This client must therefore hold the root certificate of the CA that issued the gateway's certificate (see Fig 9.4).

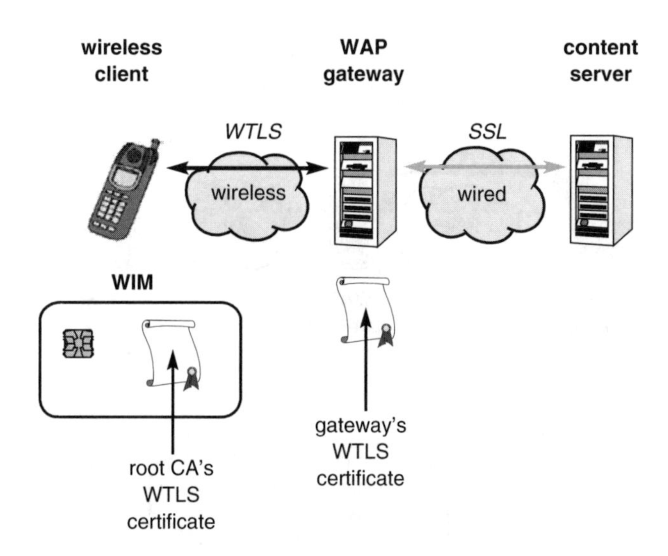

Fig 9.4 Securing the wireless connection using WTLS.

Ultimately the root certificate will be stored in a wireless identity module (WIM), as described in section 9.3.4. The major difference from the SSL, fixed Internet link, is the limited capacity of the wireless client to store and validate certificates. This introduces the need for a new compact certificate format — the WTLS certificate [8].

In addition, the client should be able to check whether the gateway certificate has been revoked. Methods already exist in the wired world — OCSP and CRL checking. Because of the constrained wireless environment, revocation checking is not an option at the present time. Instead, short-lived WTLS certificates could be issued to the gateway [6].

9.3.4 Authentication of the Wireless Client

Many applications, for example in the financial sector, will need to be assured of the identity of the client, and none of the methods discussed so far have addressed this. Figure 9.5 shows the two methods by which this can be realised:

- WTLS client authentication between the WTLS end-points (Class 3);
- WMLScript signText digital signatures between client and content server.

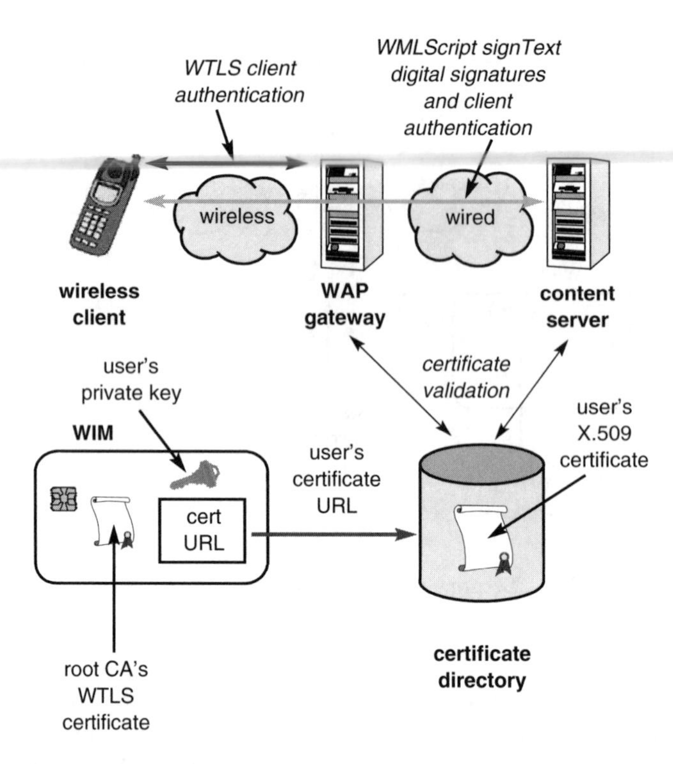

Fig 9.5 Authenticating the client and signing transactions.

Both these methods require that the client possesses a private key and certificate. The private key must be stored on the client, in this case in a tamper-resistant hardware token called a wireless identity module (WIM) [9]. This can be collocated with the existing SIM card to form the so-called 'SWIM' card. An alternative approach, which may be driven by the financial community, is to implement the WIM as a separate smart card, similar in size to a credit card.

As the client certificate is to be verified in the wired world, it must be a conventional, full-size X.509 certificate [10]. Rather than storing the full-size certificate on the WIM, a pointer to the location of the full certificate, typically a URL, is used.

The relying parties, gateways and servers, can easily access the full version from a directory of such certificates.

It is envisaged that initially the same key pair will be used for both WTLS client authentication and for digital signing at the application layer. In time, a dual key pair approach will be adopted, separating the authentication key from the signing key [6].

9.4 The Reality of Implementing a WAP PKI

The previous sections have summarised the theoretical design of the WAP PKI. This section discusses some of the key implementation issues at the time of writing.

The brief overview in section 9.3 highlights the many parties that need to work together to make a commercial WPKI solution. Each technology provider has a piece of the jigsaw, and each of the jigsaw pieces must join together to provide a complete and interoperable service. There are at least five players:

- handset manufacturers — provide compliant handsets;
- handset browser suppliers — provide compliant browsers;
- smartcard vendors — provide WIM or SWIM cards;
- WAP gateway providers — contribute conforming products;
- PKI vendors — supply components such as software to establish a certification authority, PKI portal (which is used in client registration), OCSP responder (which is used to check to see if a client certificate has been revoked), and application tool-kits (to allow end-application developers to verify a digitally signed transaction, for example).

A potentially confusing issue when attempting to assemble these constituent components into an overall solution is the seemingly loose 'mapping' between the desired security features and vendors' compatibility claims against a particular WAP release. For example, WTLS specifications appear in all the WAP releases to date, but not all handset manufacturers may have chosen to fully implement these standards. Specifically, a nominally WAP 1.1-compliant handset chosen at random from those available today has no assurance of support for WTLS Class 2.

In addition, even though an 'approved' version of the WPKI specification [6] can be obtained from the WAP Forum at the time of writing, it is not listed as being in the WAP 1.2.1 (June 2000) conformance release [11]. So even though most of the desirable security features (WTLS client authentication and WMLScript digital signatures) are in WAP 1.2.1, much of the detail as to the PKI surround is not. One important area of interest, for example, which has not been considered in this chapter, is the registration process by which end users are issued certificates. The WPKI specification assumes that full over-the-air registration is possible, whereas this is not necessarily the case today. This is encouraging many PKI vendors to develop interim solutions to address this issue.

9.5 Summary

The evolution of the WPKI is a fast-moving area, and the next few years are likely to see further dramatic advances as previous restrictions, such as network bandwidth and wireless client capabilities, become less of a constraint. It is anticipated that there will be increased take-up of desirable security features, such as complete end-to-end confidentiality and integrity at the application layer. Eventually, WAP standards are likely to converge with the W3C/IETF Internet standards, providing features such as full TCP, HTTP, XHTML and TLS at the handset.

The inclusion of a digital signature capability early in the development of WAP client devices will accelerate the widespread acceptance of PKI as a leading authentication technology. Indeed the unique nature of a wireless device in providing secure storage of a user's credentials in a mobile environment, combined with developments in short-range wireless technology such as Bluetooth, could see the mobile telephone become the ubiquitous authentication token for both the wireless and wired worlds.

References

1 WAP Forum — http://www.wapforum.org/

2 WAP Forum: '*WAP architecture specification*', (April 1998) — http://www1.wapforum.org/tech/documents/

3 Netscape: '*SSL 3.0 specification*', (November 1996) — http://home.netscape.com/eng/ssl3

4 WAP Forum: '*Wireless transport layer security specification*', (February 2000) — http://www1.wapforum.org/tech/documents/

5 WAP Forum: '*WMLScript crypto library*', (November 1999) — http://www1.wapforum.org/tech/documents/

6 WAP Forum: '*WAP public key infrastructure definition*', (April 2001) — http://www1.wapforum.org/tech/documents/

7 WAP Forum: '*WAP transport layer end-to-end security*', (June 2001) — http://www1.wapforum.org/tech/documents/

8 WAP Forum: '*WAP certificate and CRL profiles*', (May 2001) — http://www1.wapforum.org/tech/documents/

9 WAP Forum: '*WAP identity module specification*', (February 2000) — http://www1.wapforum.org/tech/documents/

10 Housley R et al, '*Internet X.509 public key infrastructure certificate and CRL profile*', RFC 2459 (January 1999) — http://www.ietf.org/rfc/

11 WAP Forum Releases — http://www.wapforum.org/what/technical.htm

10

TETRA SECURITY

D W Parkinson

10.1 Introduction

There has been an explosive growth in large-scale mobile communications systems triggered by the introduction of the total access control system (TACS) in the early 1980s. TACS was an analogue mobile telephone system, and was followed in the early 1990s by the fully digital global system for mobile communications (GSM) [1]. The launch of the new universal mobile telecommunications system (UMTS), otherwise referred to as 3G or third generation system [2], is now imminent.

In addition to these systems, we have seen the development of the digital enhanced cordless telecommunications (DECT) for cordless telephony (home, PABX), and for the professional mobile radio market the terrestrial trunked radio standard (TETRA) [3]. TETRA is becoming the system of choice for public safety organisations (police, fire, ambulance, etc) and is the basis of the BT Airwave service — a public radio safety communications service [4, 5].

As the demand for data services has grown, the general packet radio service (GPRS) has been developed to meet the needs of data users in GSM, while both TETRA and UMTS have integral support for packet data services.

All these mobile communications systems are based on open standards that have been developed within the European Telecommunications Standards Institute (ETSI) [6] and its affiliated organisations (e.g. GSM Association [1], TETRA MoU [3] and the 3rd Generation Partnership Project — 3GPP [2]).

One common characteristic of all these mobile communications systems is the very obvious fact that the connection between terminals and the network is via radio, a medium that is very easy to intercept in both a passive (eavesdropping) and active (impersonation) sense. Exploitation of this fundamental vulnerability attracted significant adverse publicity for the earlier analogue systems. This has made security an essential aspect of all subsequent systems, and each new system has built on the experience gained from the previous generation. This is exemplified by GSM, the world's first digital public mobile radio system, where the use of cryptography was introduced. Cryptography is used primarily to deliver two key services:

- a strong authentication mechanism to allow the network to check the identity of all terminals connecting to the network;
- encryption of the air interface to provide privacy for user data (telephone conversations) at least equivalent to that available on the wire-based public switched telephone network (PSTN).

The former mechanism stopped dead the massive cloning fraud that had been occurring in the previous analogue networks (this did not fix the fraud problem — the fraudsters h... ernatives such as subscription-based fraud).

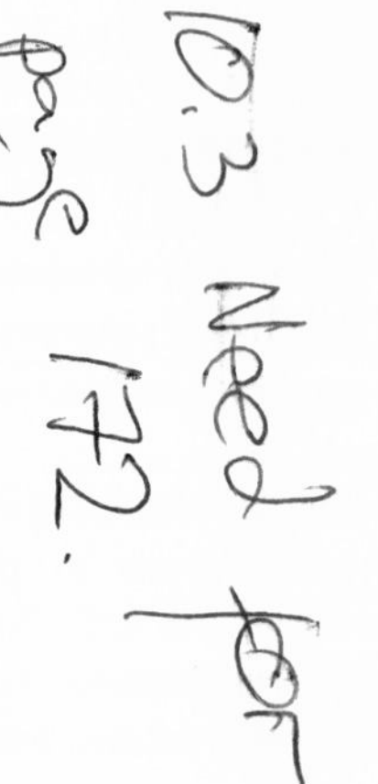

ironment

mobile station (MS) and switching and sed to designate the respective components ommon word 'terminal' is used rather than

e borne in mind that the terminal may be nge from fixed installations in vehicles, o any mobile telephone user, through to the various component parts may be rusively by an operative.

there are at least two perspectives — that r. The user's concerns primarily focus on iality of their communications, while the y to control access to the system and its ne stream is safeguarded. Luckily these exclusive.

litional concerns when considering the s within the law enforcement arena who nad or the national criminal intelligence

frequencies, and, while an encrypted them to eavesdrop on the air-interface traffic, there is still the possibility that they can gain useful information from the terminal/infrastructure signalling such as the identity of a user. Just knowing that a particular group of terminals are active in the area can be of significant benefit to criminals who can then alter their behaviour to the detriment of a particular police operation.

Also the communications of these specialist users often carry a higher government security classification than is delivered by the standard infrastructure[1]. These issues lead to additional requirements for anonymity and enhanced confidentiality. The former is supported in the standard TETRA security mechanisms while the latter is provided through an additional service, end-to-end encryption, which is covered later in this chapter.

10.4 The TETRA Security Model

In a similar way to the existing GSM and evolving 3G systems, the fundamental security service in TETRA, on which all else is founded, is strong authentication which is based on proof of knowledge of a secret that is shared between a terminal and the infrastructure's authentication centre (AuC). The authentication is both explicit and implicit. Following an explicit authentication exchange, there is an on-going implicit authentication of a terminal as the cipher keys that are then used for the up-link air-interface encryption are derived from that initial authentication exchange.

It is worth noting that in the TETRA context all authentication is between the terminal and the infrastructure, never terminal to terminal, and not directly infrastructure to user[2].

The authentication process is based on standard symmetric key cryptography. No use is made of either asymmetric (public-key) cryptography or digital certificates in the system. As has been proven in GSM, the symmetric approach is secure, efficient and easily manageable. Figure 10.1 illustrates the basic layering of the model.

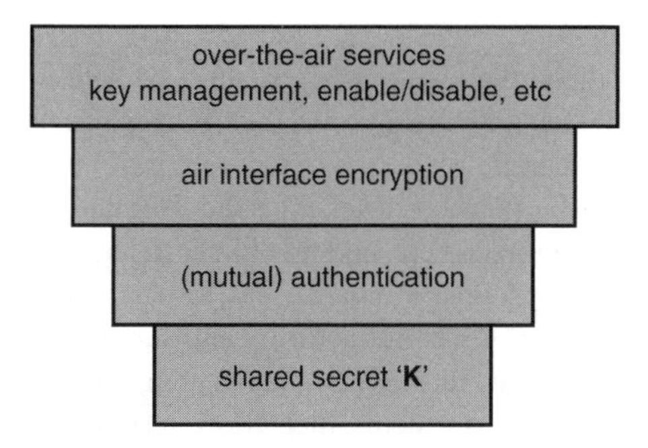

Fig 10.1 The TETRA security model.

The full description of TETRA security can be found on the ETSI Web site [7]. This document is still subject to change requests as a result of the deployment of

[1] The BT Airwave system is suitable for unclassified traffic and traffic classified RESTRICTED.

[2] In a manner similar to existing GSM handsets, a user may enter some form of PIN to 'unlock' the terminal, but it is the terminal that is authenticated by the infrastructure.

real-world TETRA systems. The changes are generally enhancements (such as group-based OTAR covered in section 10.4.2) or clarifications to deal with the complexities of the interaction of security classes (see section 10.4.3) when the system handles a fall-back from one class to a lower class on system failure.

Essentially within TETRA there are a set of security mechanisms which are used to deliver a set of security services. The next sections look first at the basic mechanisms, and then at the services that are delivered using them.

10.4.1 Security Mechanisms

10.4.1.1 Identity

In a TETRA system there are a number of unique pieces of information that may be used to identify and/or address terminals:

- each piece of terminal equipment includes a unique identifier — the TETRA equipment identity (TEI);
- the terminal equipment has an assigned individual terminal subscriber identity (ITSI) which is effectively the terminal's telephone number — it may also be programmed with one or more group identities or group short subscriber identities (GSSIs) (also referred to as talk groups);
- the terminal equipment holds the shared secret **K**[3], either directly within the terminal or in a subscriber identity module (SIM) card — **K** is effectively a secret 128-bit encryption key.

By its very nature the authentication key **K** must be held in secure memory, not be externally readable by any means, and not be replaceable other than in the factory. It must remain secret.

Ideally it should not be possible to change the TEI but the ability to do this is likely to vary between manufacturers, and it is likely to require specialist equipment. As always in real-world systems, a balance has to be struck between the security ideal and the practicalities of the manufacturing and support processes.

Operational factors may make the ITSI changeable, and GSSIs may be changed easily (there are over-the-air protocols that allow terminals to be readily moved between talkgroups). In an operational system it is the coupling between ITSI and **K** that is used to determine a terminal's identity.

---

[3] In practice, there are a number of ways in which the user authentication key can be derived, either directly or indirectly. For the purposes of this chapter, it will be assumed that the key is held directly by the terminal.

10.4.1.2 The TETRA Authentication Algorithms (TAAs)

These are a set of algorithms that are used as the basis of authentication and encryption key derivation. The full specification of a standard set of algorithms, TAA1, is available from the TETRA MoU [3]. The algorithms were developed for the TETRA MoU by the security algorithm group of experts (SAGE), a group within ETSI that was set up some years ago to design fit-for-purpose algorithms for various ETSI standards. The Airwave system uses the TAA1 algorithm set.

10.4.1.3 The TETRA Encryption Algorithm (TEA)

This is used to protect both user data and signalling on the air interface. TEA is a stream cipher with an 80-bit key. The full specification of a number of possible algorithms is available from the TETRA MoU, each tailored to a particular commercial objective[4]. The Airwave system uses TEA2, an algorithm that has been reserved for use within Schengen Public Safety organisations and their close friends.

10.4.2 Security Services

Based on the above mechanisms TETRA offers a number of security services that are essential in any practical system.

10.4.2.1 Authentication

The basis of all the security resides in a unique secret that is shared between a user and the AuC in the TETRA network. This secret may be (securely) embedded in the TETRA terminal or in a SIM card. The authentication protocol (which is based on proving that the other party has knowledge of this secret) supports both unidirectional and mutual authentication. In the former case the terminal verifies that it can trust the network or the network authenticates the terminal that is connecting to it. In the latter case both authenticate each other as part of the same transaction.

10.4.2.2 Encryption

Encryption is used to protect both user data and signalling on the vulnerable air interface. It also provides implicit authentication until such time as a further explicit authentication is carried out.

---

[4] In particular, this relates to the ability to readily export TETRA equipment to target markets in specific parts of the world.

10.4.2.3 Anonymity

Anonymity is supported through the encryption of the signalling channel on the air interface.

10.4.2.4 Over-the-Air Rekeying (OTAR)

In any system that uses cryptography (assuming good algorithms and protocols have been chosen), the fundamental security of the system lies in the cryptographic keys and the way in which they are generated, distributed, used and protected. There can be significant downstream operational costs in the day-to-day management of these keys and any consequential equipment-handling requirements. TETRA minimises these costs through the use of the OTAR service. Being able to deliver key material securely in this manner removes the need for terminals to be returned at intervals to a central point to be filled with fresh keys. It also enables the system to handle scheduled and unscheduled key changes without any disruption to the users.

10.4.2.5 Closed User Groups (CUGs)

CUGs allow the system traffic to be partitioned between the various users of the system, i.e. a broadcast transmission to one particular group of users (the Royal Parks Police, for instance) can be isolated from all other users in the system. The primary mechanism for this is the address attached to the transmission. Terminals will only unmute if they recognise the talkgroup address that accompanies the transmission. While this is a standard approach that has been used in trunked systems for many years, TETRA also offers a much stronger separation mechanism — cryptographic separation. By using an explicit group encryption key, users are assured that any group transmission can only be decrypted by a terminal loaded with that particular key.

10.4.2.6 Terminal Enable/Disable

This feature is important in any commercial system and especially so in a system with as many potential users as the Airwave service where dealing with the problem of lost or stolen terminals will be a daily occurrence. Any working terminal in the wrong hands is a breach of security[5], and the ability to remotely disable terminals (and subsequently re-enable them if they are recovered) is an essential requirement. This needs to be done in a secure manner, otherwise the feature turns into a liability as it provides an attacker with the opportunity to carry out a denial of service attack. TETRA uses the appropriate security mechanisms to safeguard this feature.

---

[5] In any trunked system, an active terminal will hold a copy of the current system-wide keys and is able to passively monitor all transmissions to the groups of which it is a member. Disabling the terminal's entry in the user database (as is done with GSM systems) is not sufficient.

10.4.3 Security Classes in TETRA

While in its original concept TETRA was designed as a secure system, commercial pressures resulted in mandatory security features becoming optional. As a result, the basic level of security offered by practical TETRA systems is defined by its 'class'. The mapping of class to security feature is shown in Table 10.1.

Table 10.1 Summary of security features in TETRA by class.

Class	Authentication	OTAR	Encryption	Enable-disable	End-to-end
1	O	—	—	O	O
2	O	O	M	O	O
3	M	M	M	O†	O

Note: M = mandatory; O = optional; — = does not apply; † = recommended.

The BT Airwave system is a Class 3 system, and supports the optional end-to-end encrypted communications.

10.5 Direct Mode Operation

While in the ideal world there would be sufficient base stations installed to ensure that radio coverage by the infrastructure extended to anywhere a TETRA terminal might be used, in practice this is not the case. For example, mountain rescue teams may occasionally have to operate in remote areas and might have to recover a person from a deep crevasse.

Also, for operational reasons, groups of users may wish to operate independently of the infrastructure. To support these situations TETRA supports direct mode operation (DMO) where terminals communicate directly with each other and not via the infrastructure.

When operating in DMO, as one would expect, the full set of security services is not available. For example, explicit terminal-to-terminal authentication is not possible as a terminal's individual secret **K** is only shared with the authentication centre, not with other terminals.

The air interface in DMO is still encrypted, however, by using a pre-loaded key, a static cipher key (see section 10.6), and there is implicit authentication as communication is only possible between terminals that have been loaded with the same key material.

10.6 Cryptography

10.6.1 Authentication

Authentication uses a standard challenge/response where, without revealing that secret, one party proves to the other that they know the secret that is shared between them. For efficiency the authentication is actually performed between the terminal and the local base station, not the central authentication centre, but this has a side effect of how authentication challenges from the terminal can be handled without the base-station having to know the shared secret. It should be noted that terminal authentication is easier as the AuC can provide triplets to the base station of {challenge, expected response, resultant derived ciper key} and does not have to reveal the secret.

To deal with this problem the AuC provides the base-station with a session key and a random seed from which it was derived using the secret **K** and a non-reversible algorithm (i.e. knowledge of the random seed and the session key does not reveal anything about **K**). The base-station uses the session key, random challenge and appropriate algorithm from the TAA algorithm set to derive the matching response and derived cipher key (see below). The random seed is sent to the terminal as part of the protocol exchange so that it can derive the same session key from its copy of **K**.

The authentication exchange may be initiated by either the terminal or the infrastructure, and the protocols support both unidirectional and mutual authentication.

Authentication of the terminal is essential to the system as it is used for the following purposes:

- controlling the access of the terminal to the network and its services;
- derivation of unique session key(s) to protect user traffic and network signalling.

Authentication of the network by the terminal is essential as it ensures:

- that the user is connected to the genuine TETRA system;
- that any control messages, such as 'disable terminal', are genuine and not a 'denial of service' type of attack;
- that unique session key(s) are derived to protect the user traffic and network signalling.

10.6.2 Key Material

TETRA systems use a range of key material for various purposes and these are covered below. Some are derived dynamically and some are loaded into the terminal either manually or over-the-air. Of these keys some are used directly and some in a modified form.

- Shared secret

  The shared secret **K** is unique to a particular terminal and is manually loaded into the terminal.

- Derived cipher key

  The derived cipher key (DCK) is derived during the authentication procedure. This is unique to the terminal and is used to protect uplink communications (from terminal to the network) and any individual downlink communications. It is also used to encrypt key material that is delivered over-the-air to the terminal. As it is unique to the terminal, it also provides implicit authentication whenever it is used, until such time as a further explicit authentication is carried out.

- Common cipher key

  The common cipher key (CCK) is generated within the infrastructure and delivered over-the-air to terminals protected by their DCKs. The CCK is common to a single location area (LA) or may be shared across a number of LAs. It is used to protect all downlink communications intended for multiple terminals in that LA. If the communication is intended for all terminals, it is used directly, or, if intended for a specific user group, it is used to modify that group's GCK (see below).

- Group cipher key

  The group cipher key (GCK) is generated within the infrastructure and is delivered over-the-air to terminals protected by their DCKs. The GCK is unique to a particular group of users and is used to provide cryptographic separation between such groups. It is never used directly but is first modified by the CCK of the location area to produce the modified GCK.

- Modified GCK

  The modified group cipher key (MGCK) is used to protect all downlink communications intended for a specific closed user group.

- Static cipher key

  Static cipher keys (SCKs) are generated within the infrastructure and are delivered over-the-air to terminals protected by the terminal's DCK. TETRA supports up to 32 SCKs. These keys are static in the sense that they are fixed keys that are used in their native form until such time as they are replaced. Their use is implementation dependent, and in the Airwave system 30 keys are reserved for use in DMO, while the remaining two keys are used as fall-back keys. The fall-back keys are used should there be a problem in the infrastructure (for example, with the AuC) forcing it to drop back to function as a class 2 system.

In addition, there are two keys specifically associated with key management. These are used to 'seal' the traffic keys before they are delivered by OTAR.

- Session key for OTAR

  The session key for OTAR (SKO) is used to seal a traffic key that is to be downloaded to an individual terminal.

- Group SKO

  The group session key for OTAR (GSKO) is used to seal a traffic key that is to be downloaded to a group of terminals.

10.7 End-to-End Encryption

10.7.1 Overview

In a standard TETRA system the encryption is only applied to the data over the radio path. It remains in the clear within the infrastructure and requires a degree of trust in the infrastructure and with all those involved in its operation. An end-to-end encrypted voice service operates between terminals without any intervention by the infrastructure other than its role as a bit carrier. It removes the need for a user to trust the network to maintain the confidentiality of the data during transit. All that is required of the network is that it acts as a transparent pipe and delivers the same sequence of bits to the receiving terminal. This is illustrated in Fig 10.2.

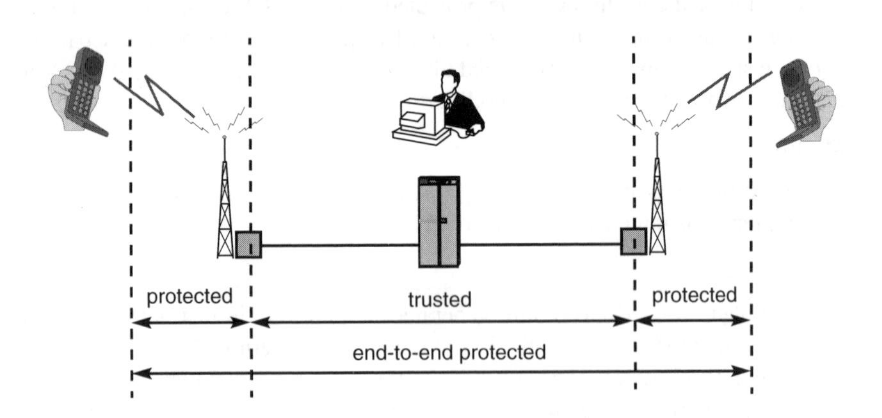

Fig 10.2 End-to-end encryption.

The TETRA standard is designed to support an end-to-end encrypted voice service. It offers the necessary transparency along with a small amount of signalling support that allows end-to-end encrypted terminals to easily achieve and maintain cryptographic synchronisation.

It should be noted that this service operates directly on the speech frames as they pass to/from the standard TETRA transport layer in the terminal. This end-to-end service adds an additional level of confidentiality to the speech traffic and is an enhancement to the system. Other than (theoretically) making redundant the air-interface confidentiality of voice traffic, such a service does not replace any of the many other standard TETRA security mechanisms.

In fact the standard mechanisms are left unaltered should an end-to-end encrypted terminal be used. It should be remembered that the air-interface encryption is also protecting the signalling between the terminal and the infrastructure, not just the voice traffic.

As was noted earlier, an essential element of any cryptographic system is the approach to key management. The TETRA standard recognises this in that it reserves a special identifier in the short data service (SDS)[6] for end-to-end encryption key management messages.

The writers of the TETRA standard recognised that the end-to-end encrypted voice service is a specialised service, and one where the potential public safety users would have strong views and policy that would be driven by individual national security requirements. Thus the standard addressed the fundamental infrastructure issues (transparency and a mechanism to assist synchronisation) leaving a high degree of flexibility in the implementation. In fact, assuming the infrastructure supplier meets the TETRA specification, implementation of an end-to-end service is purely a terminal issue.

10.7.2 The TETRA MoU End-to-End Encryption Framework

Several years ago the TETRA Security and Fraud Prevention Group (SFPG) [8] was established. The role of this group was to deal specifically with security-related issues to do with TETRA. The group has produced a number of recommendations in relation to how practical TETRA systems should be operated. In particular, they have addressed issues, such as key distribution [9], where there has been obvious benefit in having a common approach in multi-vendor environments.

The SFPG recognised that, in the area of an end-to-end encrypted service, many users, while having particular requirements over the cryptography used, may have

[6] SDS is the TETRA equivalent to the GSM short message service (SMS).

no wish to specify the rest of the end-to-end system, including the key management. As a result, it produced TETRA MoU SFPG Recommendation 02 [10]. This recommendation fully specifies all that is required for an end-to-end service other than the detail of the cryptographic algorithms. These are treated as black-box functions.

The framework has been designed to be adaptable to a range of security policies, with the flexibility being achieved through a number of simple operational choices. (e.g. setting the number of user groups).

The BT Airwave end-to-end encrypted service is based on Recommendation 02.

Recommendation 01, on key distribution, is downloadable from the SFPG Web site [8] while Recommendation 02 (end-to-end encryption) is available to TETRA MoU members on application to the SFPG secretary whose contact details are given on the SFPG Web site [8].

10.7.3 End-to-End Encryption Algorithms

SFPG Recommendation 02 is written around four black-box cryptographic functions designated E1 to E4. Those users with the necessary expertise may define how these are realised using algorithm(s) of their own choice. The only constraint is that the algorithm(s) have to fit within the broad parameters of functions E1 to E4.

For those users who are content to follow a public standard, the recommendation includes an appendix that shows how these cryptographic functions can be realised using the IDEA algorithm. Therefore the body of the recommendation together with the appendix forms the complete specification for a standard TETRA end-to-end encrypted voice service. On this basis the TETRA MoU has established a licence agreement with ASCOM (the owners of IDEA) covering the use of IDEA in this context (see the TETRA SFPG [8] for details).

The BT Airwave end-to-end encrypted voice service uses algorithms designed by CESG [11], the UK's National Technical Authority for the official use of cryptography.

10.7.4 Key Management

The approach to key management in Recommendation 02 was designed with operational costs in mind. It is similar to the standard TETRA security in that each terminal holds an embedded secret — the key encryption key (KEK). This secret has to be directly loaded into the terminal. It is expected that the KEK would be unique to a terminal as this minimises the management overhead if the terminal is lost or there is a suspected compromise of its key. Thereafter all other keys may be loaded over-the-air using the SDS service.

The next key in the hierarchy is the group encryption key (GEK). The GEK is expected to be shared across a group of terminals that are managed in an identical fashion. By using a shared key it allows a very bandwidth-efficient broadcast method to be used for the distribution of traffic encryption keys (TEKs). The TEKs are the keys that are used to encrypt the voice traffic.

In a similar fashion to the standard service, the available management messages include commands that allow end-to-end keys to be deleted from the terminal (end-to-end 'stun and kill').

One other thing should be noted — by using the SDS service as the key distribution mechanism, the end-to-end key management centre does not have to be part of the infrastructure. SDS messages may be sent from any TETRA terminal and thus the key management centre may be considered to lie entirely within the user domain and not be part of the BT Airwave infrastructure.

10.8 TETRA Security in Practice

Let us consider the practicalities of getting a TETRA terminal into service on an existing TETRA infrastructure. The infrastructure will have its own security policy and secure management procedures. As part of the security architecture appropriate measures will have been taken to ensure that an adequate level of protection is afforded to its databases and elements such as the authentication centre.

The early stages of the process given here are described in SFPG Recommendation 01 which gives a standard file format that may be used for the communication of information between the various parties involved in the process.

The terminal is manufactured and is programmed with an identity — the TEI. When the terminal is procured it will have its secret **K** programmed into it. This may be done as part of the manufacturing process, or at a separate secure location using a specialised programming tool. The value **K** may have been created by the manufacturer using an approved method, or may have been supplied to the manufacturer by the security authority of the organisation procuring the terminal. At this stage we have a TEI/K pairing. This information is delivered in a secure manner to the infrastructure AuC.

The terminal then passes to the service point at which the ITSI is programmed. ITSIs, like telephone numbers, are issued by a central authority. The service point will have a block(s) of ITSIs that have been allocated to it, and it in turn will assign ITSIs to terminals in accordance with the system policy. The service point, having programmed the terminal with an ITSI, will forward the TEI/ITSI pairing to the AuC.

At this point the AuC database can be populated with the relevant information (primarily the ITSI/**K** pair), but the terminal will not yet be made active.

The terminal is then assigned to a user (or group of users if not personal issue). At this stage it may be preprogrammed with the necessary talkgroups. Also information relating to the user such as group membership(s) and privileges (the ability to make calls via a PSTN gateway, for instance) are entered into the system management databases and the terminal is enabled on the AuC.

When the terminal is first switched on, it authenticates with the infrastructure. At this point the terminal identity passes in clear as the terminal contains no active key material. Once the terminal is authenticated the terminal ends up with its DCK. Missing key material can be delivered by OTAR, and standard TETRA management commands can be used to complete the configuration of the terminal. On completion the terminal is live on the system. When further re-authentications are performed the terminal identity is protected by encryption.

When the terminal is switched off (e.g. at the end of a shift) it is optional as to whether a terminal deletes all of its dynamic key material or retains it in secure memory. When the terminal is switched on again, if all the key material was retained and is still valid it will be used to protect the terminal's identity during authentication. If no valid keys are present the terminal's identity is transmitted in the clear.

In general usage, the operational key material in the terminal is updated using group-based OTAR. From a practical perspective a group approach to key updates offers a very significant saving in signalling bandwidth and operational convenience without unduly undermining the overall security of the system.

Whenever the terminal is out of infrastructure for a period of time, either because it has been switched off or because it is being used in DMO, there is always the possibility it will miss a change of keys. To deal with this situation in a simple manner the infrastructure regularly broadcasts the identities (serial numbers) of the keys in use. If the terminal finds it is holding out-of-date keys it will automatically request up-to-date keys from the infrastructure.

Finally, if a terminal is lost or stolen it can be explicitly stunned or killed. This feature, which is not present in conventional public telephony mobile systems such as GSM, is necessary due to the different nature of TETRA. In GSM the only key material present in the terminal is that pertaining to the user. In TETRA the terminal holds key material (such as the CCK, GCK and SCKs) that is shared with other users. It is important that these shared keys can be erased remotely. It also prevents a misplaced terminal from passively monitoring traffic. Just disabling a terminal in the AuC (the GSM approach) is not sufficient.

10.9 Beyond the Standard

In TETRA, as with GSM, because the standard authentication is between the terminal and the infrastructure, it says nothing about the user of a terminal. In a public safety system this could be regarded as equivalent, if terminals are personal issue items. However, where terminals are fixed in vehicles, this is obviously not so. In any case, some form of personal logon adds an important additional layer of security. The Airwave system supports such a scheme, but, as with all things, its realisation is not a straightforward matter in the context of a real-world operational system with its multiplicity of terminal types and user behaviour.

10.10 Summary

This chapter has described the way in which security has been designed into TETRA, and has placed it in the context of users of the BT Airwave system. As with any system, the security requirements do not remain static, but continue to evolve over time to meet the needs of real-world users. Also the valuable experience gained in the design of the TETRA Security Architecture will be (and has been) carried forward into the design of future ETSI mobile standards.

References

1 GSM Association — http://www.gsmworld.com/

2 Third Generation Partnership Project — http://www.3gpp.org/

3 TETRA MoU — http://www.tetramou.com/

4 BT Airwave service — http://www.bt.com/airwave/

5 Tattersall, P. R.: '*Professional mobile radio — the BT Airwave service*', in Clapton, A. J.: '*Future mobile networks: 3G and beyond*', The Institution of Electrical Engineers, London, pp 221-232 (2001).

6 European Telecommunications Standards Institute — http://www.etsi.org/

7 ETSI: 'ETS 300 392-7 terrestrial trunked radio (TETRA); Voice plus Data (V+D); Part 7: Security', — http://www.etsi.org/

8 TETRA SFPG — http://www.tetramou.com/sfpg/index.html

9 SFPG Recommendation 01 — Key Distribution — http://www.tetramou.com/sfpg/

10 SFPG Recommendation 02 — End-to-End Encryption — http://www.tetramou.com/sfpg/

11 Communications-Electronics Security Group — http://www.cesg.gov.uk/

11

FIREWALLS — EVOLVE OR DIE

D J Gooch, S D Hubbard, M W Moore and J Hill

11.1 Introduction

The phenomenal growth of the Internet continues apace. IP networks are cheap, highly flexible, endlessly upgradable, and ubiquitous. Everyone uses them; nobody can now afford not to. Businesses have been quick to recognise the value of the Internet as a source of information, and as a new medium for communicating with customers. However, they have been slightly more reluctant to use the Internet as a means of interconnecting their own internal networks, despite the obvious cost savings offered. Accustomed for so long to being isolated from the outside world by firewalls, and to the connection of remote sites using private networks, the idea of trusting private packets to a public transport medium like the Internet has been slow to catch on. Although virtual private network (VPN) technologies, which allow the creation of secure overlays on untrusted networks, have been in existence for some years, it is only recently that implementations and standards have reached a level of maturity appropriate to the commercial world.

The automotive and finance industries have been among the first to realise the potential benefits of VPNs. Motor manufacturers in the USA quickly recognised IP networks as a more efficient way of communicating with partners, suppliers, customers, and attracting new customers. The result was the Automotive Network Exchange (ANX), the world's first commercial secure IP network [1]. The ANX now extends into Europe and Asia/Pacific, and has even begun to offer its services to customers outside the automotive industry. The banking and finance sectors, driven by the same commercial needs and the rising costs of private wires, built their own secure IP networks. Vendors quickly recognised the commercial opportunities of IP VPN. The Internet Engineering Task Force (IETF) strove towards common standards acceptable to all. The technology matured. Now more and more businesses facing similar commercial drivers are prepared to link their networks to the Internet.

Security mechanisms are not part of the standard TCP/IP protocol suite that is currently used to carry data on IP Version 4 networks. Therefore, in order to protect data and systems on a private network connected to the Internet, and to secure communications between that network and the outside world, additional security measures need to be taken.

Traditionally, information security is broken down into fundamental component services:

- confidentiality — to ensure, normally by cryptographic means, that data is not disclosed to unauthorised parties;
- integrity — to ensure, again normally by cryptographic means, that data has not been altered in any way;
- availability — to ensure that legitimate users are not, either by accident or by malicious intervention, denied access to services to which they are entitled;

and optionally:

- accountability — to ensure that the actions of all users of systems and networks can provably and reproducibly be tracked, so they can be held accountable for those actions; accountability relies in turn on the presence of adequate access controls and authorisation mechanisms.

These services apply equally to data in storage and in transit. Maintaining accountability can be particularly costly if carried to extremes. The decision as to the level at which each of these services should be pitched must naturally be based on a proper cost-benefit analysis. Underlying all these security services is authentication. There must be a reliable means of identifying the creator and the source of data.

In the context of commercial IP VPNs, we can afford to be a little less formal. There are three underlying requirements that must be satisfied by any IP VPN technology:

- packets carried over the VPN must be logically isolated from other traffic on the network (particularly when it comes to addressing);
- there must be appropriate support for data security services, i.e. authentication, access control and authorisation, data integrity, data confidentiality, accountability — for Internet VPNs, availability is a major concern;
- there must be appropriate support for quality of service guarantees — guarantees of average and peak bandwidth, and latency, should at least match those offered by more traditional frame relay, asynchronous transfer mode (ATM) or leased-line VPN services.

11.2 The Traditional Security Solution — Firewalls

Firewalls are regarded as the primary defence mechanism for connecting private IP networks to the Internet. There are various types [2], but firewalls essentially do a single job. The reason for their presence is to enforce an organisation's security policy. The security policy for the firewall states what sort of connections are allowed through it. The firewall then makes an access control decision for each IP packet or connection that it sees, based on the data for that packet or connection in the security policy. This decision is a simple 'Yes' (allow the packet or connection to pass) or 'No' (drop the packet or connection), although the rules that govern individual firewall implementations are often very complex. The firewall controls access at the application and transport layers, often using application layer (contextual) information to make its access control decisions. Unless the firewall is proxying connections at the application level, the only items of information it has about a particular connection are the source and destination IP addresses, and the port numbers where TCP and UDP based protocols are being used.

A firewall is there to stop unwanted traffic from the Internet entering the protected network, by virtue of prohibiting such connections in its security policy. In a similar way the firewall's security policy can also control traffic going out on to the Internet. By its nature a firewall is a perimeter-based security mechanism. It has no control over what occurs on the internal network. It simply screens any connections made between the inside and the outside.

Typically, firewalls are set up to allow outgoing connections but to disallow any incoming ones. In this case it could be argued that the firewall provides a level of authentication for internal traffic in that it prevents any externally originated packets from getting on to the internal network.

Market research shows that the firewall market is continuing to grow as more and more businesses link their networks to the Internet. The global firewall market is expected to grow from \$1.1B in 2000, to \$4.2B in 2005 [3]. In the past, firewall vendors have targeted large business customers but they are now releasing products that are scaled down to appeal to small office or home office (SOHO) users and individual users. The adoption of broadband, 'always-on' Internet connectivity to small offices and home users is also encouraging this trend.

However, business models are changing, and an increasing number of companies are concluding that perimeter-based firewalls are too restrictive.

11.3 The Changing Business Model and Role of the Firewall

The firewall is the traditional favourite of the security manager, as it presents an easy way of controlling traffic into and out of the domain. To the security manager, the firewall is a single point where security and audit functions can be applied. However, to developers and application users firewalls are often considered an inhibitor that prevents them from getting real work done.

Frequently, this is because firewalls are configured to block traffic from applications that a user wants to use, e.g. NetMeeting and other chatroom applications. Similarly developers frequently wish to use protocols disallowed by the firewall rule set. In many companies the role of the IT manager also includes responsibility for network security, although such is their workload, that the security role is more reactive in response to reported problems, rather than proactive. Therefore, developers and users are likely to have much more time and motivation to attempt to circumvent firewall restrictions.

At this point it is worth considering what firewalls cannot protect against. The most obvious example is an attack that does not pass through the firewall, e.g. one directed via a computer with a modem that is connected to the corporate network. Another example is a network-based attack that is launched from inside the firewall (this is where about 70% of all computer attacks originate according to a Computer Security Institute [4] study in 1998). Furthermore, firewalls cannot protect against tunnelling over allowed protocols, and this is discussed in section 11.4. Although this should be clear, the point is well made that a business should not focus security at a single point — the firewall should form only a part of a comprehensive security architecture.

The perimeter security model, where a wall is built around the company network, has reached the point where it is no longer viable. Corporate break-ups, mergers and federations of companies, are commonplace. The ability to rapidly tailor the firewall's policy to accommodate this dynamic environment is becoming increasingly important. Yet security needs to be pervasive and not obstructive to business. Businesses also have the need to compartmentalise within the network as well as between networks.

Businesses are making more use of home working, and this trend is being accelerated with the deployment of broadband technologies such as ADSL to the home. ADSL is providing much higher data rates, when compared with PSTN and ISDN. However, ADSL's always-on, or always-available connectivity is coupled with concerns over the security implication that always-on means always-vulnerable, especially in the case of ADSL hosts that are connected directly via an ISP, instead of via a business intranet. Companies are also outsourcing parts of their business and making more use of application service providers (ASPs). The net impact of this is an increase in the number of connections and the amount of traffic that needs to pass through the firewall. This, in turn, increases the size of the firewall required and complicates its management.

11.4 Protocol Tunnelling

An example of how easy it is to bypass firewalls is presented by protocol tunnelling which encapsulates one protocol inside another [5]. Tunnelling is a general technique which can be used to carry a protocol across a foreign network. If a perimeter firewall only knows the source and destination addresses and ports for each connection, then, by using a protocol tunnel, it can be fooled into allowing a protocol to pass, that it is supposed to be blocking. Tunnelling is often used to join two isolated networks with a private bridge across a public network, forming a

VPN. Most commonly, IP traffic is encrypted and encapsulated in a TCP stream, which is carried across the public Internet between two remote sites. IPsec is such an example, and this is discussed in section 11.5.

Any protocol can be exploited for tunnelling. The only requirement is that the protocol is permitted by any firewall that sits between the tunnel end-points. Protocols like SMTP and HTTP generally satisfy this requirement. Others, like ICMP-ECHO (the 'ping' protocol), are also allowed by many configurations.

A protocol tunnel can turn an application layer protocol (such as HTTP, or SMTP) into a transport layer protocol. This technique is used by media streaming software based on Quicktime [6] or Real Networks' Real Player [7] to make the streaming protocols they use work better with perimeter firewalls. This can make it very hard for the firewall to reason about the traffic passing through it, and several tools are available that exploit this fact (e.g. Gopher [8]). The obvious problem caused by tools such as this is that the firewall policy no longer dictates the overall security policy. Although users must take a conscious decision to invoke applications like these, it is ultimately the users who can decide which particular protocols will cross the firewall.

The protocol tunnelling example shown in Fig 11.1, follows the basic HTTP protocol to set up connections. A simple program is used at the client and server ends to read and write messages, and manage the tunnels (the server can manage tunnels to multiple clients simultaneously). The client and server queue messages at each end of the tunnel:

- if the client has messages to send, it makes an HTTP POST request to the server, with the messages included within the body of the request;
- otherwise, if the client has no messages to send, it makes an HTTP GET request to the server;
- if the server has any messages to send, they are included within the body of the response.

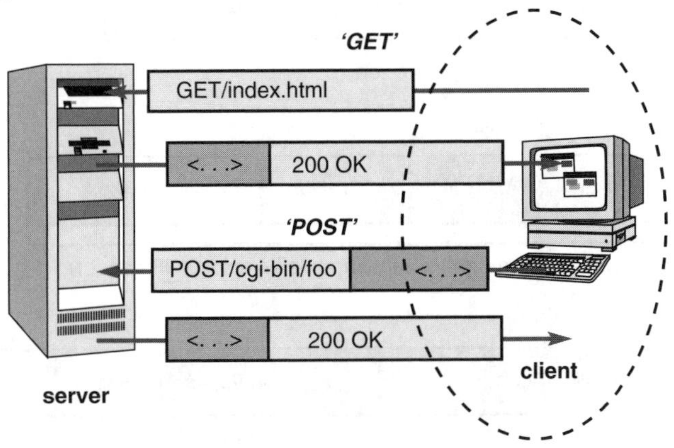

Fig 11.1 HTTP protocol tunnelling.

Therefore perimeter controls go only part way to providing security for a private network connecting to the Internet. A firewall can be bypassed if a host on the inside can be compromised. A network of insecure systems cannot have its security policy enforced by a perimeter firewall alone — but the problem is not the ability of the firewall, it is the security of the systems behind the firewall. A strategy of 'defence in depth' is critical in preventing attacks such as the one just described. This term has gained popularity recently and it is often used to refer to host-based detection of viruses or intrusions — but true in-depth defences should protect against these attacks too. The next section introduces the IPsec network security, and discusses where this can provide greater protection for the network.

11.5 IPsec

IPsec is a suite of IETF protocols (defined in RFC 2401-2412) that provide security services for traffic at the IP layer. IPsec uses cryptography to provide authentication, integrity, confidentiality and replay protection at the packet level, and is a technology which permits the setting up of secure virtual circuits over IP networks.

IPsec can protect traffic between hosts, between network security gateways (e.g. routers or firewalls), or between hosts and security gateways. There are two IPsec protocols, the encapsulating security payload (ESP), and the authenticating header (AH); these are IP protocols 50 and 51 respectively. AH (see Fig 11.2) provides authentication of received packets, data integrity and anti-replay protection. ESP (see Fig 11.3) also provides these services (although the scope of the authentication coverage differs), and adds optional data confidentiality. AH and ESP can be used alone or combined, typically using ESP encryption with AH authentication. For both AH and ESP the strength of the security services is dependent on the

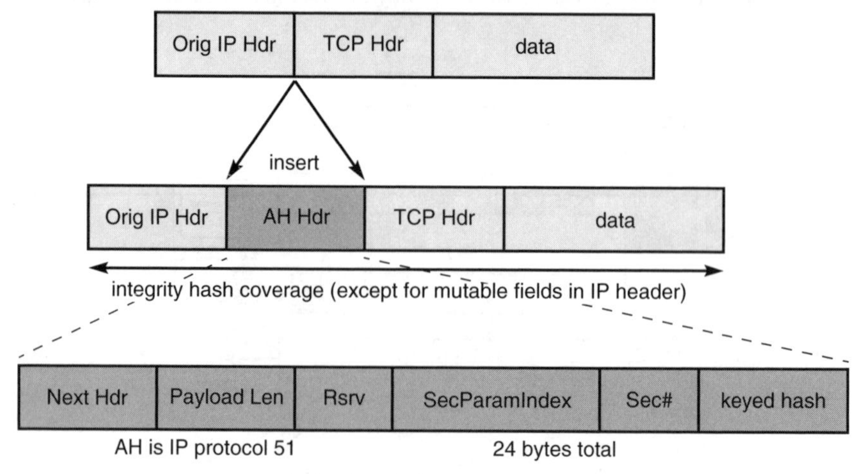

Fig 11.2 AH protected packet.

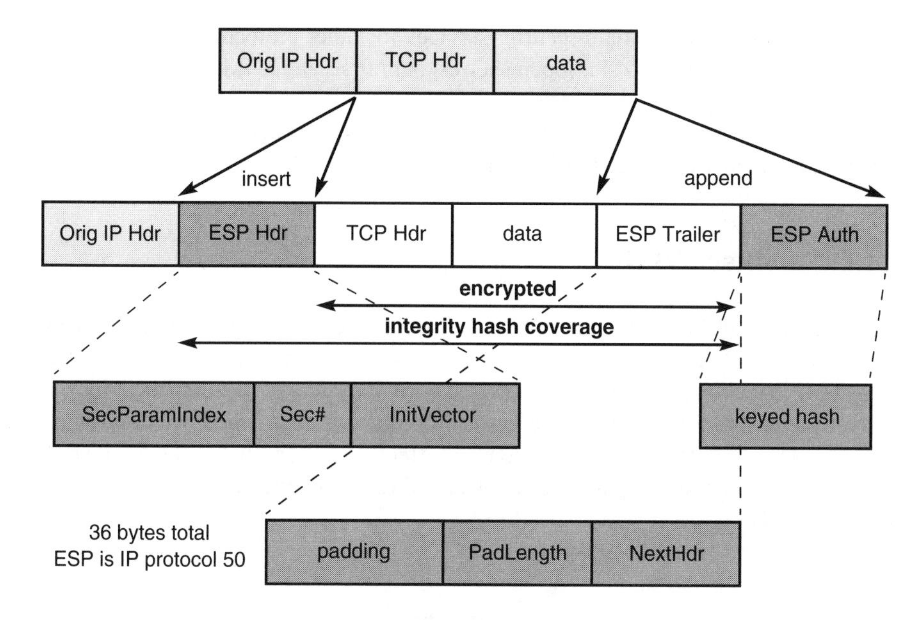

Fig 11.3 ESP protected packet.

cryptographic algorithms used, e.g. Triple DES (Data Encryption Standard [9]) is the current favourite for ESP. However, plans are already afoot to adopt the Advanced Encryption Standard (AES) [10].

There are also two modes of IPsec, known as transport and tunnel modes. Tunnel mode, which is currently more common, is designed to secure traffic between network security gateways, where it can provide security services on behalf of other network entities, e.g. hosts behind gateways which are connected to form VPNs. Transport mode is only designed to secure traffic from host to host, i.e. where the cryptographic end-point is also the communications end-point. At the expense of some extra processing, tunnel mode can also be used to protect end-to-end communications. Both AH and ESP protocols can be used in either mode.

Prior to securing IP packets with IPsec, a security association (SA) must exist, and these can be created either manually or dynamically. The Internet key exchange (IKE) protocol is used to negotiate SAs dynamically on behalf of IPsec — IKE provides strong authentication of the identities of the communicating hosts. Typically, IKE authentication uses X.509 digital certificates from a separate public key infrastructure (PKI), although it can also use keys that have previously been set up via an out-of-band mechanism, so-called 'pre-shared secrets'. The IPsec SA that results from IKE comprises an authenticated key, and agreed security services (derived from the security policy). Thus IKE provides the means for the two end-points to exchange keys securely, over an insecure network. IKE is not only used for

IPsec — it can also negotiate security services for other protocols that need them, e.g. the routing protocol OSPF (open shortest path first) and SNMP (simple network management protocol) v3.

A detailed description of the various IPsec transforms is beyond the scope of this chapter; the reader is referred to the IETF IPsec RFCs [11].

11.6 IPsec VPNs

In the context of this chapter, a VPN is defined as the 'emulation of a private wide area network (WAN) facility using IP facilities' (including the public Internet, or private IP backbones). The use of VPNs is growing quickly as companies look to reduce the cost in long-distance charges for remote users and leased lines for remote offices. The key to the construction of effective virtual networks is flexible management, in provisioning the VPN and throughout the rest of its life cycle (see Fig 11.4).

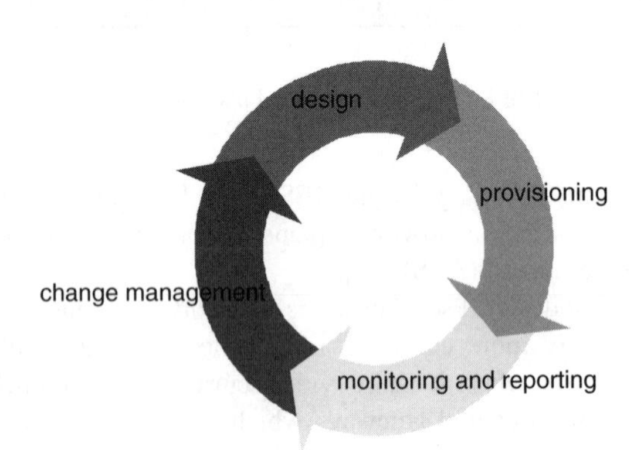

Fig 11.4 VPN management life cycle.

IP quality of service (QoS) mechanisms can provide for data separation, albeit without providing strong security. QoS and security have therefore tended to become muddled. Consequently the term virtual private network has become a bit of a misnomer, perpetuated by marketing.

Only networks providing cryptographically strong separation of user data can be said to be truly private.

VPNs based on IPsec are capable of carrying traffic securely over many types of shared networks, including the Internet. IPsec VPNs offer the ability to solve corporate communication problems while also cutting costs; applications include remote access for road warriors, as well as secure intranets and extranets.

11.7 Extending Security to the Desktop

It has been shown that IPsec can be implemented directly on security gateways, such as routers and firewalls, and end hosts. There are two ways of implementing IPsec in an end host. It can be either a bump-in-the-stack (BITS), where the IP stack is modified to support IPsec natively by inserting a shim between the network and data link layers (as has been done in Windows 2000 and XP). Alternatively, when a software solution may not be desirable, or indeed possible (an IP-capable printer, say), a bump-in-the-wire (BITW) can be used; this is typically an external hardware device that performs IPsec processing. BITW has the potential to offer superior solutions to BITS implementations because of the inherent speed advantages of dedicated cryptographic processing hardware that forms part of a BITW (Fig 11.5). A BITW can provide the user with transparent security equivalent to a personal firewall which may be managed from the network rather than the end host. A BITW based on open source software also allows cryptographic plug-ins and can be host independent, easy to install, and potentially cheap for a large-scale uptake.

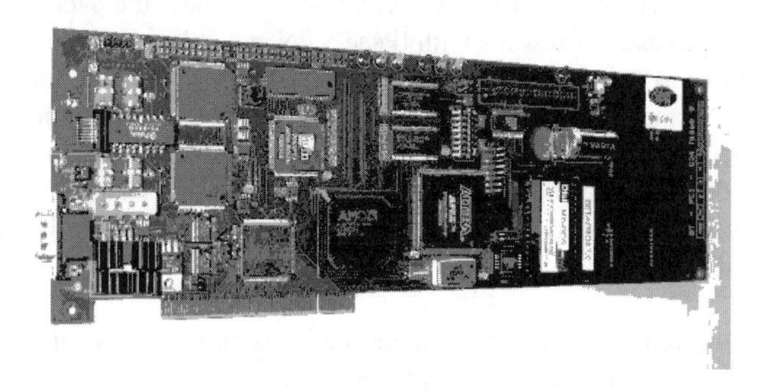

Fig 11.5 BITW prototype developed by BTexact Technologies.

When applied to end entities, both BITS and BITW can extend security services end-to-end — but when is it necessary to extend security to the end host? One answer may be where several businesses share the same building or site with a common communications infrastructure. In this example, businesses may not be able to afford their own security gateway; it may therefore be a requirement to use IPsec to protect the confidentiality of traffic from potential eavesdropping. Remote access home workers and road warriors may also require security to extend to the end host (although this may terminate at the home network's security gateway).

One possible disadvantage in using IPsec to provide end-to-end security, is network address translation (NAT) which is used to connect private intranets to the Internet. NAT is applied at the boundary of the private network to map intranet private addresses to addresses that are publicly routable. However, if the entire packet has already been secured with IPsec, NAT will corrupt the packet (the IETF are currently assessing potential solutions to this problem).

End-to-end security enables VPNs to be built without a perimeter, enabling networks to become virtualised. In this scenario security is devolved to end hosts, and, with communications secured with IPsec, traditional firewalls are not necessary. Despite this, the threat of 'denial of service' attacks against the whole network remains, and firewalls may be used to filter this traffic. As firewall functionality becomes pervasive throughout the network, it becomes vital that all the parts of the firewall are managed in a secure manner. Section 11.8 discusses some of the options for these next-generation networks.

11.8 Firewall and VPN Policy Management

One of the major functions of the traditional perimeter firewall is to act as an enforcement point for the organisation's network access security policy. Essentially, the firewall has to make a decision, based on the information that it can gather about the packet and its current security policy, whether to allow the packet to pass or to drop it, or, in the case of firewalls with IPsec VPN capabilities, to process the packet with IPsec prior to forwarding it. This decision is easy to make at the perimeter of the network, because all the packets traversing the firewall are either entering or leaving the trusted domain. When the firewall moves into the network the decision becomes harder, because of the multiplicity of routes that a packet can take. As the firewall becomes more and more distributed, the policy that controls the firewall's processing of packets becomes more complicated and correspondingly harder to manage.

Some of the more enlightened vendors have recognised that this is a problem, and have come up with solutions that allow management of large federations of firewalls. Typical examples of such solutions are those from Cisco [12] and Checkpoint [13] and from VPNet in the VPN space [14]. However, today's policy management solutions work mainly with a single vendor's products. Partially, this is as the result of a commercial decision by the vendors, who are naturally keen to lock their customers into a single compatible product line. Unfortunately, this philosophy is not really appropriate for the real world, for as soon as another vendor's products come into the picture the situation changes. Typically, the circumstances in which this happens arise when companies merge their network infrastructures. The banking, telecommunications and Internet sectors are particularly prone to this phenomenon. It is also a particular problem for companies such as BT, who sell network management services.

Following a merger between companies deploying firewalls from different vendors, the firewall manager is left with the problem of how to manage the overall security policy using multiple vendors' products for implementation. One or two technical solutions are starting to appear which make a start at solving this problem.

An example is SolSoft's Net Partitioner product [15]. A typical view of the Net Partitioner interface is shown in Fig 11.6.

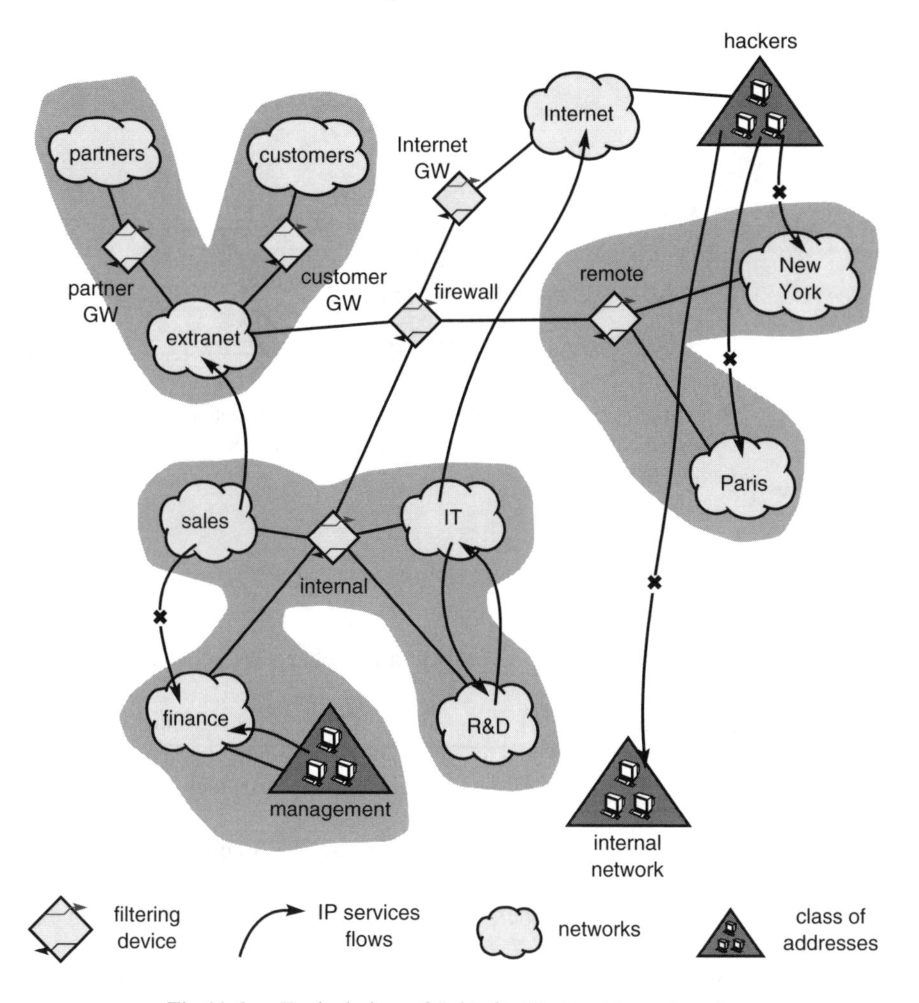

Fig 11.6 Typical view of SolSoft's Net Partitioner interface.

To use Net Partitioner, the firewall administrator needs to know the detailed topology and traffic flows on the network. Once the network has been defined, the administrator can then decide which services are allowed between the various parts of the network. This completes stage 1 of the process shown in Fig 11.7.

The product then converts the graphical policy description into a set of access control lists (ACLs) (stage 2), which it downloads on to each policy enforcement point in the network (stage 3). These devices may be routers, firewalls, switches or other devices — Net Partitioner supports a range from various vendors. Finally, it

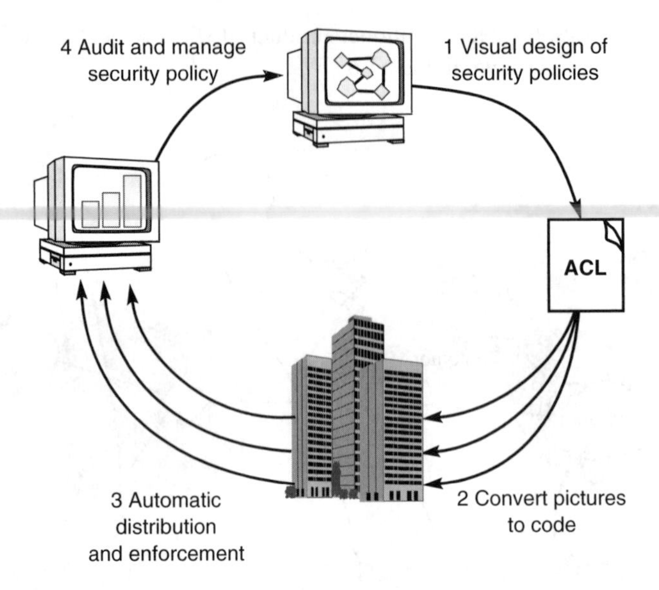

Fig 11.7 Net Partitioner's policy management stages.

includes a set of audit and monitoring tools to allow changes to the network to be checked (stage 4).

The TBD Networks tool-set [16] works in a similar way to Net Partitioner, but the devices it can manage are confined to VPN gateways and clients. The user firstly has to define the network in terms of the devices and those to which each device is allowed to talk.

Once this has been done the TBD tool-set can manage VPN connections between diverse gateways and between gateways and clients. In many ways, the TBD tool-set is more limited than Net Partitioner, because it concentrates only on provisioning and managing VPNs.

Using either of these tools demonstrates that there are several key requirements for policy management, whether it is of a firewall or a VPN.

- Security policies should be easy to define

  The user of the system should be able to concentrate on the policy, and not on the technology that implements the policy. The two commercial products mentioned above take a proprietary approach to policy definition — they have their own policy definition engines. Some work has been done on writing a generic policy description language, at AT&T Labs and other places. The resulting trust management system, called Keynote, is generic in nature, but has been applied to IPsec VPNs [17].

- Security policies should be easy to distribute

  The next major requirement is that security policies must be easily distributed throughout the network. There are essentially three ways of accomplishing this:

  — the client may poll the central manager for the necessary information, which is sent on demand;

  — the manager may push the policy data to its clients when required;

  — the manager deposits the policy in a central repository, such as a directory, which is interrogated by the clients.

  Some implementations may choose to use a combination of methods of policy distribution. The current version of Netlock (a software IPsec VPN solution [18]) uses a combination of the first and second methods. Solsoft NP's and TBD Networks' solutions utilise only the second method. Microsoft's Windows 2000, with its Active Directory, utilises the third method.

  Of these alternatives, having the manager send the policy to each and every agent using some form of 'push' protocol is probably the easiest to implement, but because the manager has to be involved in every communication between itself and the clients, it scales poorly. Likewise, having the client poll the manager every so often for policy updates also makes for a less scalable solution, and has the additional complexity of the polling protocol as well. Of course, actual implementations are more complex than this — the Netlock client, for example, polls the manager for policy updates at system boot time, but once it has successfully asked for and received the latest policy from the manager, nothing changes until it receives an update pushed to it from the manager. The use of a directory to store the policy and to give it out when required is probably the best method of distributing security policy, because:

  — the manager does not have to be involved in every communication between itself and the clients that it manages, which improves the performance of the manager;

  — directories may be distributed and duplicated across a number of servers, which improves reliability, performance and scalability of the policy distribution mechanism;

  — security policies should be easy to change.

Security policies are not static objects; they evolve as the requirements of their users change. Therefore, it is necessary that a state-of-the-art security policy management system must make it easy for policies to be changed. The policy change mechanism should be the same as the policy distribution mechanism, with the added requirement for some way of alerting a client of a change in policy. There are several ways of tackling this. Netlock, for instance, includes the ability for the central manager to push policy updates to individual agents as soon as any policy change that affects them is made. Another way is to make policy changes centrally and to alert individual clients that such updates are pending — this approach is taken by Checkpoint.

- Security policies should be easy to audit

  Security policies may be very complicated. Additionally, in a managed service provider (MSP) scenario, an MSP is likely to have to manage secure network infrastructures belonging to multiple customers. The MSP will wish to keep tabs on changes to each customer's VPN. Likewise the customers themselves will wish to ensure that no-one has made any unauthorised changes to their infrastructure. Therefore it is of paramount importance that any deployable management system will have high-quality auditing abilities, enabling a customer to see, for example, all changes that have been made to their network.

11.9 The Future

The earlier sections have shown that the function of the firewall is to keep unwanted traffic out of the network being protected, and indeed this should continue to be the firewall's role. How this is achieved is what is likely to change. In our future world, secure virtual networks will be established over existing infrastructures. These IPsec VPNs will have no fixed perimeter, and therefore no single point to place a traditional firewall. If all connected parties communicate using IPsec, responsibility for security is devolved to the policy enforced at end hosts. However, this scenario does not render firewalls obsolete, as an attacker could still inject packets into the network. It is therefore proposed that what is really needed are firewalls distributed throughout the network.

Taking the example of a distributed denial of service (DDOS) attack, the targeted host is all too aware of the attack, and may be able to discard packets without service being adversely affected. However, too often this is not the case and valuable business can be lost. Therefore ideally, the host would want to have these packets filtered in the network prior to reaching their intended destination. The ideal point(s) being the upstream routers. If all communications are using IPsec, it would be trivial for 'router firewalls' to filter this traffic. However, IPsec connections are established using IKE, which is not itself an IPsec protocol. Hence injected IKE packets could be used to launch an attack.

Prevention of this type of attack would rely on a relatively simple enhancement to IKE to allow the host to detect and react to these packets. Similarly, a router firewall could not filter spoofed IPsec packets, which again would need to be dropped by the host. In both cases, injected packets would initially have to be dropped by the end host. Therefore, the end host would need an IPsec secured mechanism for informing the upstream router firewall of the need to filter the unwanted packets. These messages would clearly need to be understood by router firewalls from different vendors, and for these messages to propagate through the network, to ultimately filter the unwanted packets as close to the source as possible. The preventative action taken as a result of these messages would require a defined lifetime, since the host originally under attack would have no means of detecting when the attack had ceased.

11.10 Summary

This chapter has shown how the traditional role of the network firewall, as a perimeter-based device designed to protect a trusted network by assessing incoming, and even outgoing, data packets in sequence, has evolved. The need to keep unwanted traffic out of a protected network remains. It is the network itself, the variety of applications it must support, and the way data is partitioned on it that is fundamentally changing. Businesses and consumers need to be able to connect quickly, easily and securely to less trusted parties. Network security can no longer be discussed in simple 'picket fence' terms.

There is still a need for firewalls, in the traditional packet filter sense, but they need to be distributed throughout the network in question, and the mechanisms by which they decide which packets, or sequences of packets, are permissible must be refined. In any case, it is highly unlikely that the concept of the stand-alone firewall will endure. Instead, filtering functionality could better be built into commonly used network devices — routers, hubs, even host network stacks themselves. The firewall is likely to become a distributed mechanism for detecting and barring data packets which, by accident or malicious design, have been introduced into the 'normal' sequences.

Much of the security functionality originally offered by traditional firewalls is now better offered by IPsec VPN overlays. The ability to secure data packets from one end of a connection to the other allows us to transcend the old connectivity barriers. With proper IP security, your data can be secure anywhere. You can connect quickly and securely to merchants, partners, customers, colleagues across any network, trusted or untrusted. Security becomes much more a question of good connection and policy management.

There has been no mention here of intrusion detection systems (IDS). These may be thought of as network or host-based systems designed to monitor data flows and search for unusual patterns that may be the result of deliberate intrusion, or accidental malfunction. A proper discussion of IDS is outside the scope of this chapter. However, many feel that distributed IDS will perform the essential complementary security monitoring function that is missing from a security architecture based on firewalls and IPsec. In the opinion of the authors, the management at least of intrusion detection systems should sensibly be kept separate from IPsec and firewall management. The latter are really there to ensure that policy is created, distributed, enforced, and can be audited. Monitoring and enforcement should be separately managed.

Unfortunately, IDS technologies are far less mature than IPsec. The original promise of versions that 'learn' new attack patterns without having to be updated, possibly using techniques from the world of artificial intelligence, as opposed to signature-based systems that have continuously to be updated, has not been fulfilled. You cannot trust wholly to IDS, but every little helps.

References

1 Automotive Network Exchange — http://www.anxo.com/

2 Hubbard, S. D. and Sager, J. C.: '*Firewalling the 'Net*'', in Sim, S. and Davies, J. (Eds.): '*The Internet and beyond*', Chapman & Hall, London, pp 153-178 (1998).

3 Datamonitor — http://www.datamonitor.com/

4 Computer Security Institute — http://www.gocsi.com/

5 Hill J: 'Bypassing firewalls: tools and techniques', FIRST Conference (March 2000).

6 Apple Quicktime — http://www.apple.com/quicktime/

7 Real Player — http://www.realnetworks.com/

8 Gopher — http://wiscinfo.doit.wisc.edu/cbt/ww4w30/gopher/gopher2.html

9 Data Encryption Standard — http://csrc.nist.gov/publications/fips/fips46-3/fips46-3.pdf

10 Advanced Encryption Standard — http://csrc.nist.gov/publications/drafts/dfips-AES.pdf

11 IETF — http://www.ietf.org/html.charters/ipsec-charter.html

12 Cisco — http://www.cisco.com/

13 Checkpoint — http://www.checkpoint.com/

14 VPNet — http://www.vpnet.com/

15 SolSoft — http://ww.solsoft.com/products/

16 TBD Networks — http://www.tbdnetworks.com/product.html

17 Keynote — http://www.crypto.com/trustmgt/kn.html

18 Netlock — http://www.netlock.com/product.html

12

THE IGNITE MANAGED FIREWALL AND VPN SERVICE

G Shorrock and C Awdry

12.1 Introduction

The Ignite Managed Firewall service is a fully managed platform providing high-quality gateway security and Internet VPNs via IPsec (Internet protocol security) tunnels. Gateway security allows for policing the boundary between the corporate network and the Internet. It prevents access to the corporate network from the Internet (keeps the hackers out), and it controls employees' access to the Internet.

IPsec tunnels provide the ability to establish secure virtual connections across un-trusted networks such as the Internet and can be used to supplement or replace traditional private networks with minimal incremental cost. In addition, customers can choose from a range of value-added features, such as URL filtering and virus protection, which, when combined with the firewall capability, offer a comprehensive solution to gateway security.

Although IP is becoming the ubiquitous standard for public and private data networks, public IP networks (i.e. the Internet) are still inherently open and insecure. The Ignite Managed Firewall service allows enterprises to interconnect public and private IP networks to gain the benefits of connection to the Internet while allowing them to have control over who is able to connect between the two and what kinds of IP traffic can be transferred.

Using Ignite Managed Firewall with the IPsec [1] VPN option, traffic sent over the Internet can be encrypted. This allows an enterprise to extend its private IP network (or intranet) to its other sites or to build an extranet with other companies while benefiting from the advantages of cost and flexibility offered by the Internet.

BT's seamless service offers consistent end-to-end control of security through the provision of the Ignite Managed Firewall service at customer sites worldwide [2].

There are three types of IP service that are referred to as 'VPNs'.

- VPNs over the Internet with no encryption

  Some remote access solutions are described as 'VPNs', but, although there is user authentication and often some form of tunnelling, there is usually no encryption, and therefore security is limited.

- VPNs over the Internet with encryption

  These are services, like Ignite Managed Firewall, which use the Internet as a transport mechanism and encrypt data passing between gateways or between a gateway and remote users. With the relaxation in export laws covering encryption, it is now possible to deploy 3 DES (Data Encryption Standard [3]) on a global basis, thus offering highly secure VPN capabilities via the Internet.

- VPNs over private networks

  These are services, like BT Ignite's IP Select, which use a private IP network (MPLS-based), shared between a number of customers but managed by a single network operator (unlike the Internet). These solutions can offer class of service (CoS) and quality of service (QoS) guarantees as the traffic remains within the domain of one network operator and the security afforded by traffic separation is comparable with traditional unencrypted solutions such as frame relay.

VPNs in the context of The Ignite Managed Firewall service and this chapter refer to encrypted IPsec VPNs over the Internet.

12.2 The Ignite Managed Firewall Service

The Ignite Managed Firewall service is a collection of best-of-breed hardware and software products providing a global, fully managed, firewall and IP VPN service. Ignite Managed Firewall gives customers the ability to protect their internal corporate data networks from intruders when providing access to the Internet. It enables customers, through the use of encryption, to protect sensitive information against interception while travelling across the Internet.

Customers can use Ignite Managed Firewall for any or all of the following:

- protection of individual LANs by firewalling Internet connections;
- establishment of an IPsec-encrypted tunnel between their sites;
- establishment of an IPsec-encrypted tunnel between their site and third parties (e.g. distributors, suppliers);
- establishment of an IPsec-encrypted tunnel from remote and roving users to their home gateway instead of using expensive international direct dial (IDD) or long-distance calls.

Ignite Managed Firewall is based on hardware running CheckPoint FireWall-1 and Cisco routers running the IOS firewall and IPsec feature set. The firewalls protect the customer's network and internal resources from external hacking and allow the customer administrators to restrict/regulate Internet access by their users.

There are two main components to the Ignite product — a managed firewall and an IPsec VPN service.

Secure Network Operations Centres (SNOCs) actively provide advanced security services such as intrusion detection, exposure analysis and suspicious activity monitoring 24 × 7. An easy-to-use secure Web interface allows customers to view the status of their firewall, view reports and to request service changes. The service is designed to be modular and flexible — customers can choose to add options such as advanced security reporting, 24 × 7 on-site maintenance, URL filtering and virus scanning.

With the IPsec VPN option, the Ignite Managed Firewall service is targeted at customers looking for an Internet-based VPN. This allows them to set up highly secure encrypted IPsec tunnels over the Internet, supplementing expensive legacy private networks with an alternative and more scalable solution. The IPsec service allows for the creation of extranets between various organisations, as well as providing remote users with secure access to their private LANs. Figure 12.1 shows the overall network architecture.

The Ignite Managed Firewall service can simply provide firewall functionality (access control), as illustrated by site A in Fig 12.1. Adding the IPsec VPN functionality sets up encrypted tunnels between firewalls (illustrated between site B and master site, in Fig 12.1). Also, with the IPsec VPN functionality, roving users are able to access the corporate LAN by dialling into the Internet, for example using the Remote Access Service (RAS), and establishing an encrypted tunnel from the user's client PC to a firewall. As stated previously, Ignite Managed Firewall now supports both CheckPoint and Cisco firewalls with interoperability being achieved via the use of IPsec tunnels.

12.2.1 CheckPoint FireWall-1

FireWall-1 software from CheckPoint [4], the market leader in firewall software technologies, is at the core of the Ignite Managed Firewall service. FireWall-1 provides all of the following:

- grants selective network access to authorised remote and corporate users;
- authenticates network users;
- ensures the privacy and integrity of communications over untrusted, public networks, like the Internet, via IPsec;
- provides content security at the gateway to screen malicious content, such as viruses and malevolent Java/ActiveX applets;
- in conjunction with the security probes deployed on the firewall, detects attacks and raises security alarms;

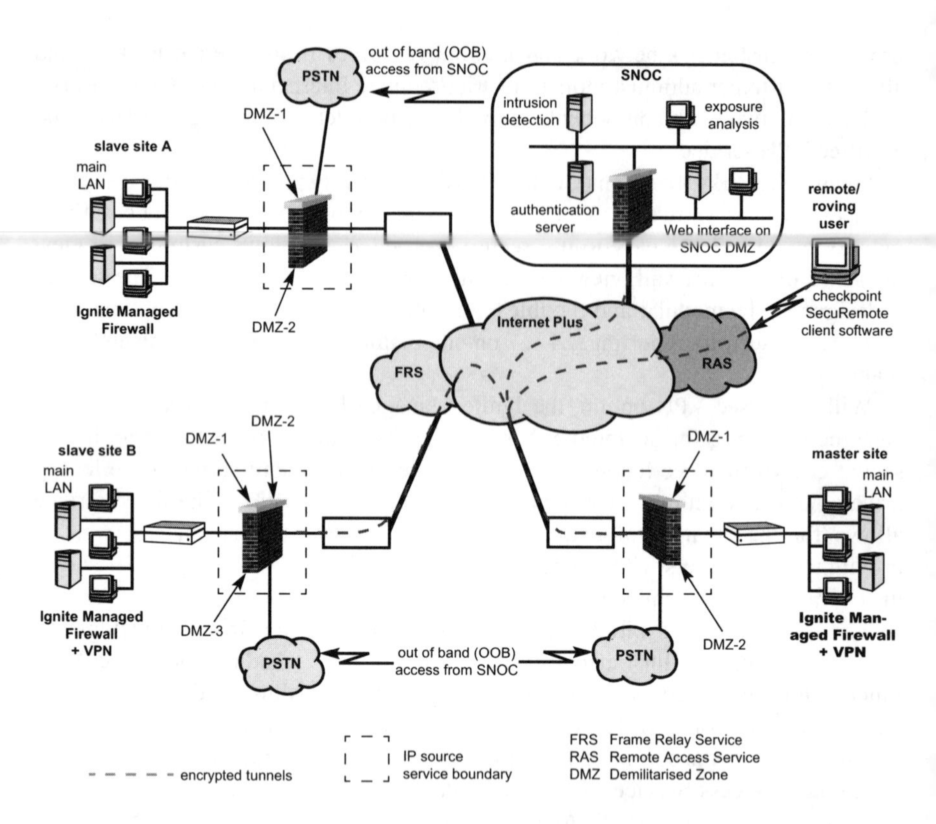

Fig 12.1 Overall network architecture.

- protects internal network addressing schemes and conserves IP addresses;
- ensures high availability to network resources and applications;
- delivers detailed logging and accounting information on all communication attempts.

Today, organisations must provide secure access to the enterprise network and its resources for a growing number of remote locations, mobile workers and telecommuters. Organisations need a way to validate the authenticity of the user or client machine attempting to make a network connection. FireWall-1 supports a wide range of authentication schemes from user name and passwords to X.509 digital certificates.

The Ignite Managed Firewall service enables enterprises to define and enforce a single, comprehensive security policy while providing fully transparent connectivity. FireWall-1 is based upon stateful inspection architecture. Stateful inspection provides the highest level of security possible by incorporating communication- and application-derived state and context information that is stored and updated dynamically. This provides cumulative data against which subsequent communication attempts can be evaluated. Stateful inspection provides full application-layer awareness without requiring a separate proxy for every service to be secured.

12.2.2 Cisco Firewall Feature Set

The Ignite Managed Firewall Cisco-based service offers a cost-effective router-based managed firewall and VPN service for those sites where a CheckPoint firewall is inappropriate. A typical scenario would be where large sites use the CheckPoint-based firewall with IPsec tunnels to smaller sites using Cisco routers. The model of router used depends on the required connection size, number of tunnels and the interface requirements. The service supports several different types of WAN connections — leased lines, frame relay, Ethernet and ISDN LAN dial-up.

12.2.3 Customer Visibility and Control of Ignite Managed Firewall

A firewall is only as good as its implementation. In today's dynamic world of Internet access, it is easy to make mistakes during the implementation process. The advantage of Ignite Managed Firewall is that it is a fully managed service, and therefore Concert's team of security experts can advise on firewall configuration and apply updates as necessary. However, the customer decides on their own security policies based on the needs of their organisation. The customer retains real-time visibility of firewall operation and can request service changes from using a simple secure Web interface which supports strong authentication and encryption.

Ignite Managed Firewall customers use the Secure Web Interface (Fig 12.2) for:

- customer management — to access their organisation's details;
- inventory management — to access details of site, contact, equipment or service;
- service requests — to request a change to the security policies;
- reports — to view reports relating to their organisation and/or site.

12.2.4 Security Management

The Ignite Managed Firewall service has developed a number of sophisticated tools for managing the security aspects of the customer's firewall. The firewalls are monitored on a 24 × 7 basis for changes to the physical configuration, the logs are monitored for suspicious activity, and a regular security scan is carried out looking for weakness in the policy or applications that are protected by the firewall.

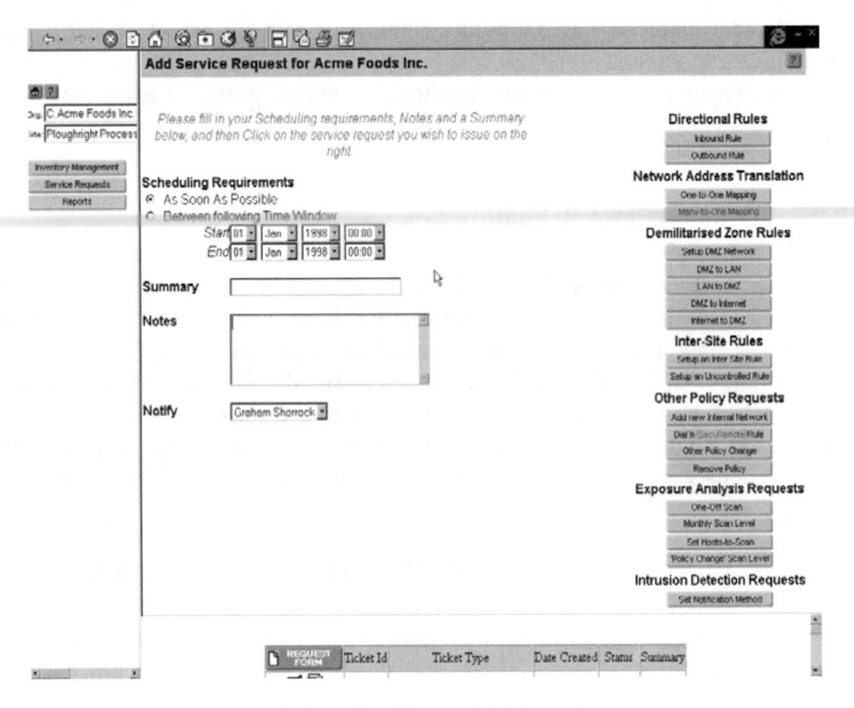

Fig 12.2 The Secure Web Interface.

12.2.4.1 Monitoring for suspicious activity

The Ignite Managed Firewall service analyses, in real time, the log events from the firewall. An examination of these events can show if a hacker is attempting to gain access to either the firewall or the customer's network. However, this analysis can be extremely difficult — gone are the days when the hacker would use a Satan scanner to look for a wide range of open ports. Today the hacker is more likely to send a single request looking for, say, a Trojan port and then move on to another IP address if unsuccessful. Only by analysing events from multiple firewalls can the stealth hacker be spotted and indeed this is the approach adopted on the Ignite Managed Firewall service. At the firewall level a probe looks for suspicious activity and, if detected, flags this to the SNOC. Here further analysis and correlation is carried out against events being reported from other firewalls and, in real time, raises security alerts to the operational staff. Figure 12.3 shows a sample intrusion detection report.

12.2.4.2 Exposure analysis

Exposure analysis is carried out using ISS Internet Scanner™ [5]. This is an automatic 'hacking' tool operated from the SNOC, which can identify weaknesses

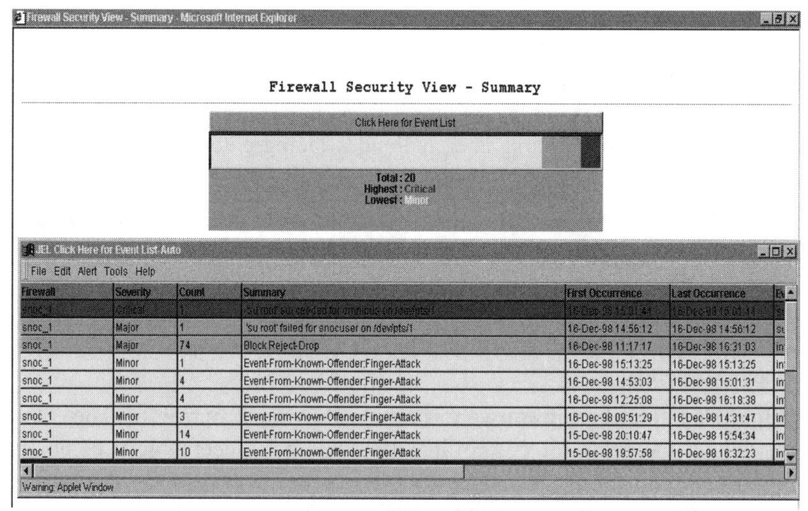

Fig 12.3 Intrusion detection report.

in the firewall security policy or hosts protected by the firewall. The resulting report (see Fig 12.4 for an example) is made available to both the customer and the operational staff so that corrective action can be taken.

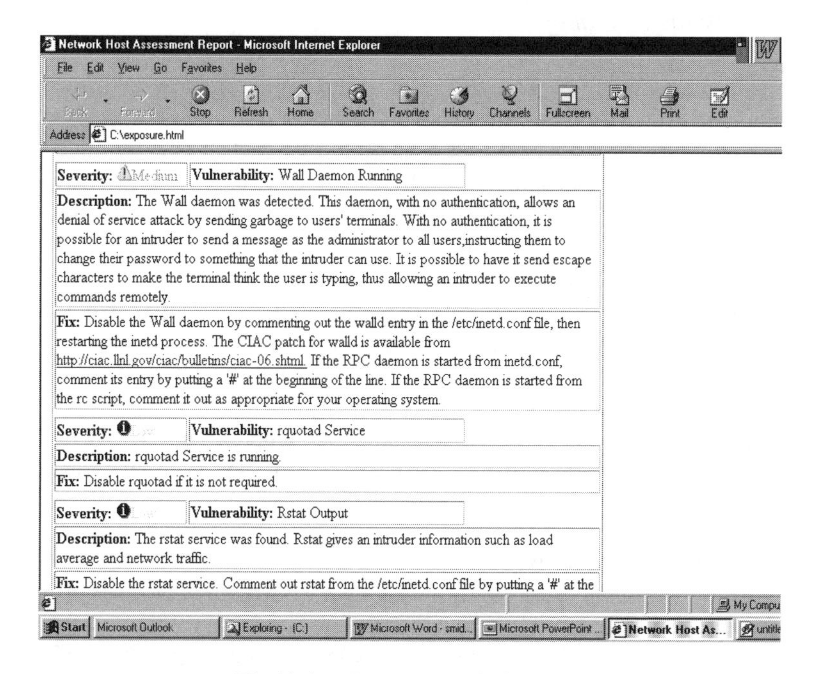

Fig 12.4 Exposure analysis report.

12.2.4.3 URL filtering

URL filtering allows a customer to control to which sites their employees have access. In surveys, up to 37% of employees have admitted surfing the Web 'constantly' during work hours. The challenges facing enterprises not using URL filtering include:

- productivity losses — as employees spend company time to visit non-business sites;
- bandwidth limitations — as leisure surfing clogs up network access;
- security threats — when harmful applets enter the corporate network from unsecured Web sites.

BT Ignite's URL filtering software is provided by SurfControl [6]. The customer can implement the rules for controlling their Internet 'acceptable use' policy by defining which classes of site (category) should be blocked and over what time period. An important benefit of the service is that the lists of URLs within each category are constantly updated. The SNOC receives the updated list of URLs from SurfControl every 2 hours, and once every 24 hours the customer's URL database is updated.

12.2.4.4 Virus scanning

Probably the most common and expensive threat to a company's IT security is contamination from viruses. Viruses have become so common that hardly a day goes by without a new virus being released on the Internet. It is therefore essential that companies protect against this threat by the use of virus scanning software. The Ignite Managed Firewall service offers virus protection at the gateway using the F-Secure FSAV software [7], which scans incoming mail for known viruses and can be applied not only to SMTP but also to HTTP and/or FTP traffic.

In order to keep the scanning software current, it is essential that the virus signatures are updated as soon as possible following the release of a new virus. This is achieved by polling the F-Secure database on an hourly basis to see if new signatures are available; if they are, the files are downloaded and the anti-virus software updated automatically. Thus within an hour (approximately) of a new virus signature becoming available, all customers will be protected from the threat.

12.2.5 High Availability

For most companies, the use of the Internet is now business critical — they simply cannot afford to have their Internet connectivity lost. For this reason many ISPs now offer resilient connectivity, but this still leaves a single point of failure in the form of the gateway. To eliminate this it is necessary to deploy a fault-tolerant solution for the gateway itself. To meet this requirement, Ignite Managed Firewall is introducing a fully redundant load-balancing firewall solution based upon software from Stonesoft [8].

Stonesoft's high availability software, Stonebeat Fullcluster, uses a single virtual IP (VIP) architecture so that a cluster of firewalls appears as a single entity. As shown in Fig 12.5, all firewalls in a cluster share the same operational IP address (VIP) (and MAC addresses).

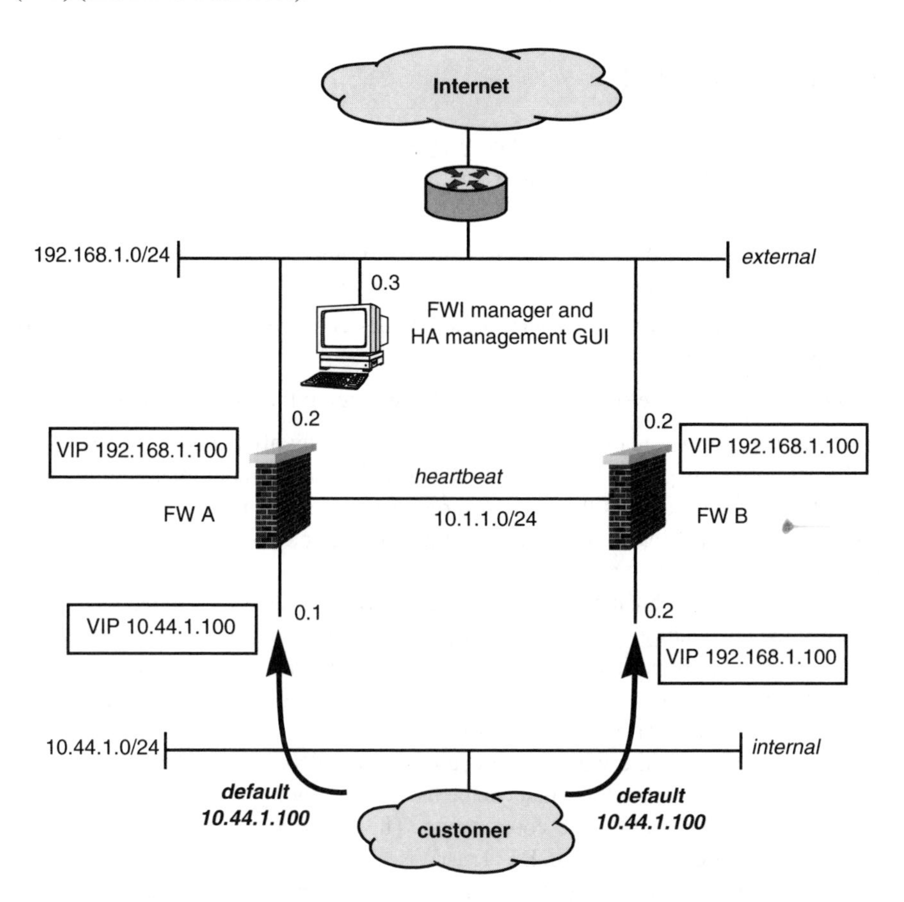

Fig 12.5 Load-balancing architecture.

With the StoneBeat solution all operational interfaces see all incoming packets. A proprietary algorithm is used to make a decision about which node in the cluster will accept the packet and this algorithm is responsible for ensuring that the return connection comes back through the same node to avoid asymmetric routing. In addition, the connection state information is regularly exchanged between firewalls in the cluster so that both firewalls know of all connections passing through the gateway. From a customer's perspective the deployment of such a solution is simple and does not involve costly re-engineering of their internal network. All the customer has to do is define the virtual IP address of the cluster as the default gateway and the Stonesoft software takes care of the rest. In the event of a failure occurring within the cluster, a test facility identifies the failure and instigates a change-over to the working system. Typically this happens in less than a second and connections that are passing through the cluster are maintained.

12.2.6 IPsec VPN

IPsec defines a set of protocols and cryptographic algorithms for creating secure IP traffic sessions between IPsec gateways. A customer can use IPsec encryption to allow:

- encrypted traffic between instances of the Ignite Managed Firewall service belonging to the same customer (extending the intranet over the WAN);
- encrypted traffic over an extranet between sites belonging to different customers or partners;
- remote users to have secure access via the Internet to their internal network.

Using the Internet as the transport platform can be highly beneficial — not only is the network truly global but the cost advantage can be considerable. Setting up an IPsec tunnel for customers that have Internet access represents a small incremental cost, when compared with the cost of an international private circuit or frame relay link.

Another area that businesses can find attractive is to use dial connections to the Internet for connecting their smaller sites. Many Internet service providers (ISPs) now offer flat fee unlimited calls to connect to the Internet — combining this with an IPsec tunnel can generate a VPN solution at huge cost savings when compared with traditional dial or private circuit offerings.

A critical area for the widespread use of IPsec VPNs is the establishment of a PKI (public key infrastructure) that can be used for providing the certificates that are required for authenticating the connections. The Ignite Managed Firewall service has leveraged the capability of BT TrustWise [9] to provide this capability in a highly secure manner. The use of certificates as opposed to pre-shared secrets provides a highly scalable solution and one that is cost effective in terms of the operational load generated by the management of the certificates.

12.2.7 Customer Scenarios

The Ignite Managed Firewall service has proved to be highly flexible in meeting customers' needs, a few examples of which are described below.

12.2.7.1 Basic Firewalling

Customers are increasingly deploying multiple firewalls at strategic locations where they have Internet connectivity. The trend appears to be moving away from a single enforcement point to more of a distributed approach, thus minimising the traffic load on their internal private network.

12.2.7.2 IPsec VPNs

A customer will typically consider the deployment of an IPsec VPN to meet one of a number of requirements.

- To back up an existing private connection

  Here the customer will typically run a routing protocol, such as open shortest path first (OSPF), over their private network, e.g. frame relay; in the event of a route becoming unavailable, an alternative route is defined via the Internet gateway.

  The firewall is preconfigured with the details of the remote site and the alternative route can be established as soon as the IPsec negotiation phase has been completed (usually less than a couple of seconds).

- To connect to a remote site where the cost of a traditional connection is prohibitively high

  By using the Internet as the transport platform, the costs can be minimised and, provided that careful policing of the Internet connection is maintained, good performance can be obtained.

- To connect to a trading partner in a secure manner

  Critically this solution depends upon tightly controlling the traffic to and from the partner's networks — here the combination of a firewall IPsec encryption and PKI infrastructure allows the communication to be tightly controlled. The firewall can define which devices and protocols can communicate, while encryption ensures the privacy of the data, and certificates provide the necessary degree of authentication.

  The European Automotive Exchange [10] is a good example of this where car manufacturers and their suppliers are establishing the capability to securely exchange information using IPsec tunnels.

12.2.7.3 Remote User Access

This is probably one of the fastest growing areas — by using an IPsec client, the travelling user can gain access to their intranet from a wide range of locations. The

IPsec client simply monitors the packets within the TCP/IP stack and those with a destination address of the intranet are encrypted and sent to the appropriate intranet gateway.

This approach is not limited to just the travelling user — home workers can use the same mechanism and, if connected to a DSL or cable modem, can achieve high performance at very modest cost.

12.3 Summary

The Ignite Managed Firewall product has evolved over the past three years to a fully featured security service offering a wide range of applications and capabilities. Probably the most exciting of these is the IPsec capability which, when combined with low-cost Internet connectivity from DSL technology, is set to fundamentally change the nature of private networking.

References

1 IETF IPsec — http://www.ietf.org/html.charters/ipsec-charter.html

2 BT Ignite — http://www.ignite.com/

3 Data Encryption Standard — http://csrc.nist.gov/publications/fips/fips46-3/fips46-3.pdf

4 CheckPoint FireWall — http://www.checkpoint.com/

5 Internet Security Systems — http://www.iss.net/

6 SurfControl — http://www.surfcontrol.com/

7 F-Secure — http://www.fsecure.com/

8 Stonesoft — http://www.stonesoft.com/

9 BT TrustWise — http://www.ignite.com/application-services/products/verisign/

10 European Network Exchange — http://www.enxo.com/index.html

13

INFORMATION ASSURANCE

C J Colwill, M C Todd, G P Fielder and C Natanson

13.1 Introduction

BT's Information Assurance Programme (IAP) aims to improve the state of preparedness to counter the emerging threat of malicious electronic-based attacks on our critical assets by protagonists with a high level of capability. As such it is directed to understand, identify and minimise the risk to our own enterprise and our customers' enterprises from the threat posed by the deliberate, unauthorised and systematic attack on critical information activities. Such attacks would be designed to exploit information, deny or affect service to authorised users, or to acquire, modify or corrupt data, and will be executed by capable, resourced and motivated perpetrators. Reaction to such attacks applies equally to BT's networks, the services it supports, and to its customers and the services they contract from BT, together with considerations of the communications infrastructure within the UK. This chapter will explain in more detail how the IAP operates and the benefits that it provides to BT and its customers. Detection and reaction processes to attacks are essential and the IAP is building upon the internal expertise in this area and BT's computer emergency response team (CERT).

13.2 A Brave New World — New Ways of Doing Business

The way companies conduct business today is changing rapidly as we move into a networked economy and infrastructure of 'eCommerce', 'eBusiness' and 'eGovernment', where the Internet is ever present. Business paradigms are changing and the integrity of systems and solutions and their guaranteed availability is becoming a key element of contracts. Availability will be dependent not only upon physical or engineering factors but also on the susceptibility of attack from protagonists who may use BT to target its customers. It is debatable whether information security policies are evolving at the same rate as business dynamics and technology. These policies must cater for dependencies on other organisations and the integration and interconnection of networks. The reasons for this lag will vary, e.g. inadequate understanding of the risks involved, an inability to visualise their impact, or the lack of focused responsibility for maintaining and funding new information security requirements.

BT is a very large commercial organisation with a complex web of IT systems, networks, and platforms underlying its key business processes. BT has interests world-wide through a variety of joint ventures and interconnections and provides a very wide portfolio of products and services to a broad range of residential and business customers. Local and global telecommunications markets are becoming increasingly dynamic and competitive and the environment is changing rapidly in terms of legislation and regulation. With operations on such a scale, the simple but key questions from any security and risk perspective must be: 'What is actually critical?' and 'Where should we invest in protection?' Focus and priorities are essential to apply cost-effective security countermeasures and other protection and reaction capabilities. BT's customers may have different perspectives on criticality and these need to be captured and understood as well as the 'internal' perspective to develop effective enterprise security measures.

13.3 Evolving Threats and Risks

In tandem with new ways of doing business, companies and their networks are becoming increasingly exposed to new threats and higher levels of risk. Risk assessment is moving rapidly away from traditional approaches, e.g. based on specific computer systems or buildings (i.e. primarily physical), towards new service and cross-portfolio/platform analysis (primarily electronic/virtual). According to Senate testimony in the USA, financial transactions in the USA over the Internet will rise from $8 billion in 2000 to $1.5 trillion by 2003 [1]. The need to address 'eRisks' comprehensively is being pioneered by BT's Risk and Insurance Solutions division, in conjunction with a number of US IT-based companies and insurance companies. The publication of the Turnbull Report [2] (which places responsibility for risk management upon company directors) has provided a driver for establishing security procedures throughout the business. The attacks of 11 September 2001 in the USA have also highlighted the fact that 'worst case' incidents can actually happen.

Customers are also becoming aware of the threats from electronic sources and, as their enterprise becomes critically dependent on networked systems, they are becoming more demanding in terms of the security capability and reputation of their suppliers.

13.3.1 Increasing Levels of 'Cyber' Attack

Electronic dependencies and interconnections create vulnerabilities that are being rapidly exploited by criminals. There has been substantial growth in the number of reported electronic breaches of information security over the past few years. A survey of major companies in the USA [3] revealed that 90% of the respondents had suffered a computer security breach in the last 12 months. Almost all of the Fortune 500 corporations have been penetrated electronically. In February 1999, the FBI estimated that US companies lose about $2 billion dollars a month from industrial espionage attacks [4]. Other notable confidentiality and integrity attacks on US systems (including military systems) are 'Moonlight Maze' in 1999 (believed to be from Russian sources) and 'Solar Sunrise' in 1998 (by three individual hackers). The impacts of the 'Code Red' and 'Nimda' worms in 2001 were substantial and are still being quantified. From a BT perspective, eleven million IP 'packets' per month are denied access through our firewalls.

In the UK, out of over 1000 organisations, each with an annual turnover of more than $16 million, only half had a formally approved computer security policy [5]. A study conducted by Information Week research estimates that the global cost of electronic security failures will be $1.6 trillion for the year 2000.

These reports underline the fact that high impact electronic attack is a real threat which can cause widespread disruption. It is widely believed that many high-impact attacks on commercial organisations are successful but are not reported to law enforcement agencies due to the potential impact on reputation.

13.3.2 National Infrastructure Implications

The importance of the nation's civil infrastructure (communications, utilities, industry, commerce, etc) in times of peace and war has long been appreciated. The growing dependence upon information technology systems and shared data, by the sectors which comprise this infrastructure, has led to the concept of a critical national infrastructure (CNI). In its most developed state, the CNI is a defined network of computers, databases, protected transmission links and security procedures.

BT's IAP has, from its inception, considered 'national' scale security implications in terms of impact on the UK economy, infrastructure and national security in anticipation of the establishment of some form of CNI within the UK. This approach was taken through consideration of various factors — Presidential Directives for protecting critical infrastructure in the USA, Y2K dependency considerations, dependency studies undertaken by BT with the UK Government, and, not least, the increasing dependency of Government and commerce upon BT's network infrastructure. The BT network not only carries BT traffic but is also used extensively by other operators under interconnect and capacity leasing arrangements.

The Home Secretary announced the UK CNI initiative in 1999 and BT has worked closely with Government and major players in all sectors to determine standards and requirements for information assurance and the CNI. BT is a full member of the Information Assurance Advisory Council (IAAC)[1] and works closely with the Government's National Infrastructure Security Coordination Centre (NISCC) to address the complex problem of defining key assets and dependencies in cyber terms.

13.4 Information Warfare

Information warfare (IW) has seen much hype in the press and on various Web sites, but it is true to say that, certainly taking into account CNI implications, the concept of sustainable high impact attacks provides a new, fundamental and substantial threat. In military terms, IW is receiving significant funding across the globe. The CIA also believes that many countries will consider using such attacks against another nation's public and private sectors during times of tension [6].

Such activities should not be confused with basic hacking using simple scripts or with actions by low-level criminals or others in pursuit of illegal or illicit ends. These actions will be undertaken by (or under the control and direction of) highly intelligent, competent, aware, well-equipped and motivated people; they may use the services of less-resourced individuals or groups as 'fronts', but the strategic intent is of a wider or deeper nature.

BT has therefore established the IAP to understand the emerging threats and the potential nature of these attacks and to implement appropriate protective measures.

13.5 Established Security and Risk Management Processes

BT has a well-established and effective set of information security processes which include criticality and risk analysis, for example:

- impact reviews (using the standard 'CIA' framework — see below);
- financial quantification of potential losses;
- identification of intangible impacts;
- protection and recovery strategies (based on risk or cost/benefit cases);
- business continuity;
- insurance levels and implications;
- security awareness;
- baseline IT security protection.

[1] The IAAC is a non-Government organisation based at King's College London and comprises members from academia, industry and commerce, and government. Its aim is to increase knowledge and understanding of information assurance aspects.

The 'CIA' framework addresses the three key elements of information security:

- confidentiality — ensuring that information is available to those with a business need and restricting any inappropriate access;
- integrity — ensuring that information is not inadvertently or deliberately corrupted or destroyed;
- availability — ensuring that corporate and customer information is available whenever, and wherever, required, and that, if a system fails, there is a fallback and recovery procedure to protect business operations from being compromised.

These processes have enabled security and risk assessments to be built into complete life cycles, covering both technical and procedural aspects of development, operation and withdrawal. The BT Security Evaluation and Certification Scheme (BTSECS) has been successful in this respect and is still the principal means of assessing and determining the security requirements of systems and products. Ownership of security at the product line level is crucial.

The BT Computer Emergency Response Team Co-ordination Centre (BTCERTCC) team was the first UK commercial organisation to gain international accreditation under the umbrella of the Forum of Incident Response and Security Teams (FIRST) [7]. Together with a total of eighty other global CERTs, BT forms a closed and trusted group, sharing technical expertise, resources and information. Together global CERTs fight cyber-crime and malicious activities on their intranets, the Internet and internal networks. Organisations who take security seriously have begun to see the need for vigilance on their networks twenty-four hours a day, seven days a week.

13.6 What is Different About the Information Assurance Focus?

In its widest sense, information assurance provides a means of giving customers appropriate levels of confidence in the security of systems, products and processes. The IAAC definition of IA is '... operations undertaken to protect and defend information and information systems by ensuring their availability, integrity, authentication, confidentiality, and non-repudiation ...' [8].

BT's IAP has developed from an initial project set up in 1997 to investigate the need for the company to consider risks to its operations arising from this potential threat. It has built upon the best practice established by the security and risk analysis communities within (and outside) BT. Existing risk assessment methods may, however, be limited because they rely upon historical data and knowledge of typical and 'traditional' threats (e.g. fraud) within a relatively known environment. New approaches must now be considered. A key success factor is the programme's support from the most senior levels of management and the strategic direction provided by stakeholders from across the BT group of companies.

The IAP seeks to identify threats to BT's corporate enterprise and to put in place measures to minimise the risks associated with them. The initial focus was on core network systems. This has developed into a programme of vulnerability and risk analysis studies, the results of which, together with the development of protagonist and business models, have enabled the construction of the 'Threat Model'. This model is currently in use to instigate and manage the ongoing programme of studies and the consequent remedial work and refinement of design criteria.

The IAP distinguishes itself by considering the implication of:

- high-capability attackers;
- high-impact targets;
- UK infrastructure and economic implications;
- lack of historical data and evidence;
- absence of clear threat warnings which could be used to justify investment;
- joint Government/industry initiatives.

The strategic aim of the IAP is to improve the state of preparedness within BT to counter the emerging threat of malicious electronic-based attacks on its critical assets by protagonists with a high level of capability.

BT's IA strategy is predicated upon an overall and comprehensive defence mechanism comprising four phases:

- **Protect** the network and the BT enterprise by prudent, feasible and cost-effective measures that will deny an attacker the opportunity to perpetrate an attack — this will include physical, personnel and logical measures;
- **Detect** events, recognising it may not be feasible or cost-effective to protect everything — mechanisms must exist to detect attacks, or attempted attacks, and determine their significance;
- **React** — either by standard processes, which may include conventional investigations, network recovery procedures, etc, or by the instigation of special IA processes;
- **Deter** — by establishing BT's status as a well-defended enterprise, attackers may be deflected elsewhere (where it may be easier) or will be less likely to attack since the probability of detection, identification and prosecution is high (BT has a strong reputation for the prosecution of criminals who seek to attack its assets, e.g. by theft or telecommunications fraud).

The IAP task is to demonstrate that it has processes and people in place to understand the emerging threats and to justify findings and recommendations for investment — all in the context of unquantifiable rather than quantifiable risk.

13.7 IA Threat and Risk Analysis

A specific BT IA Threat Model [9] has been developed to assess emerging threats (see Fig 13.1). This includes a major departure from traditional security and risk analyses (within a commercial organisation at least) in that an attempt is made to understand the threat from potential attackers' perspectives and not just from BT's own criticality and vulnerability perspective. For example, even if a high-impact weakness is identified, who out there may be thinking of exploiting it and are they capable of doing so and sustaining the attack?

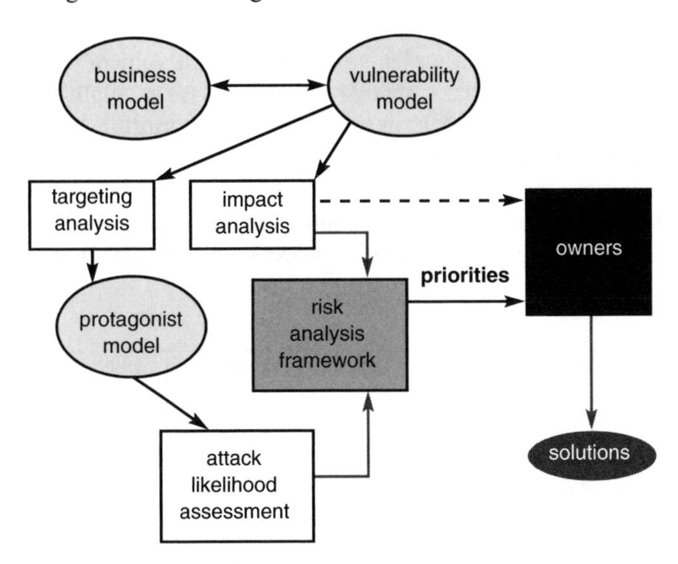

Fig 13.1 Threat Model process.

The Threat Model comprises three distinct but interrelated models.

- Business model

  This identifies critical assets such as the technology and processes underlying BT's commercial activities, including linkages and dependencies. On to this are mapped the assessments of impacts, complemented by IA specific priorities, which help to identify targets of potential interest to attackers.

- Vulnerability model

  This provides a framework for a series of IA-based rapid impact analyses designed to decompose BT's critical components to identify vulnerabilities above and beyond known weaknesses. Of particular importance is the need to challenge assumptions, e.g. the reliance on recovery options, and to challenge normal commercial considerations. Penetration tests are also performed where appropriate and the result is a set of vulnerabilities categorised in terms of impact, potential exploitation, visibility, detectability and recoverability, together with a set of recommendations for countermeasures.

- Protagonist model

  The primary aim of the model is to assess threats by considering the likelihood of malicious attack from a range of potential sources which may see BT as a target. The likelihood of attack is taken as a function of motivation, opportunity and capability.

The outputs from the models are integrated via the IA 'risk analysis framework'. Risk is taken simply as a product of impact (from the vulnerability studies) and likelihood of attack (from the protagonist model); each vulnerability is assessed against each protagonist profile. A set of 4 × 4 matrices (as exemplified in Fig 13.2) is used to determine relative risk scores to create overall priorities for action and to provide a consistent framework for making assessments over different target areas, e.g. comparing processes to systems. For risk management purposes, additional risk calculations are made when countermeasures are agreed to assess reductions in impact or likelihood — an iterative process is essential but not always straightforward.

Impact / Likelihood	Minor	Serious	Enterprise	National
Likely	7	11	14	16
Probable	4	8	12	15
Possible	2	5	9	13
Remote	1	3	6	10

Fig 13.2 Likelihood and impact matrix.

13.8 Using the IA Protagonist Model

The IA protagonist model [10] attempts to understand and differentiate potential groups of attackers by profiling differences in their motivation, opportunity and capability (see Fig 13.3).

The term 'protagonist' is used in its widest sense — BT is interested in key players who may have the opportunity and capability to perpetrate malicious electronic attacks, but whose motivation or intention of attacking is unknown. The following generic profiles are maintained:

- criminals;
- commercial competitors;

- nation states;
- terrorists;
- hackers;
- disaffected employees.

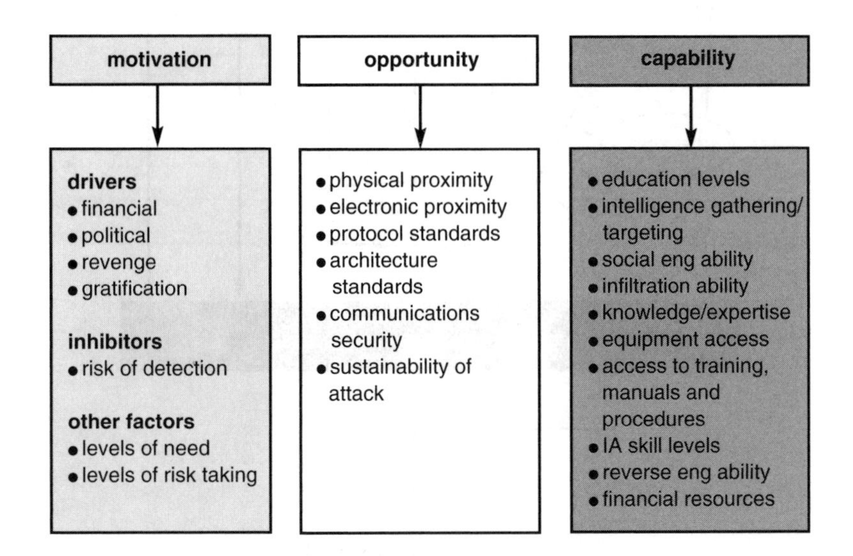

Fig 13.3 Protagonist profiling.

Each of these groups may be broken down into a hierarchy of more specific profiles if it is believed that actual identifiable groups may be targeting BT. Where necessary, interaction and co-operation/collaboration between protagonists may be modelled, e.g. to assess composite attacks and the use of agents (see Fig 13.4). Each profile may have a set of relatively 'given' scores based on the typical *modus operandi* of a group (type of attack, technology used, etc). There will also be a set of 'variable' factors when looking at BT as a target, e.g. extent of electronic connectivity and knowledge of operations.

Specific scenarios and assessments are then set up to understand the likelihood of an attack on a given target area and set of vulnerabilities. This is done because even though a weakness may exist there may be few, if any, protagonists who can exploit it — and those who could exploit it may not wish to do so. It is important to avoid assumed/implicit threats or to focus just on 'headline' attackers and a rigorous framework is provided explicitly to assess threats from other sources as well. Experience shows that significant threats and risks also come from less obvious sources.

The assessments take the form of intensive workshops during which attack scenarios are developed, varying roles and perspectives are taken and the profiles

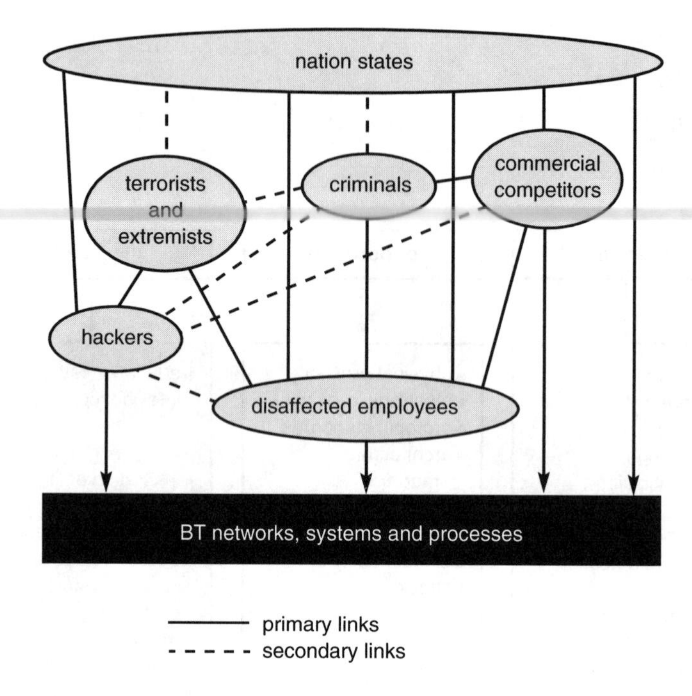

Fig 13.4 Composite attacks.

applied to specific target areas. It is important to involve both security experts and people familiar with the target area.

The sessions will usually involve much iteration and sensitivity analysis, arriving, ultimately, with consensus on the results. The workshops are supported by a facilitator and spreadsheet.

The assessment of attack scenarios must consider the electronic, physical and also the psychological domains and interactions between them. A protagonist might use social engineering to develop their attack during the targeting stage and might then use more traditional physical methods (e.g. bombing) to draw attention away from a more subtle electronic attack, which is their end game. Multiple attacks are possible within each of these three primary domains, e.g. an 'electronic' attack on our data networks could be supplemented by an associated 'electronic' attack on the PSTN designed to thwart our recovery attempts. Figure 13.5 shows a typical targeting cycle for an attack.

In addition to these considerations, the attack can take several forms, dependent on whether it is an attack on confidentiality (i.e. theft of the content of an asset), integrity (i.e. manipulation of an asset), or availability (i.e. destruction of an asset or denial of access to an asset).

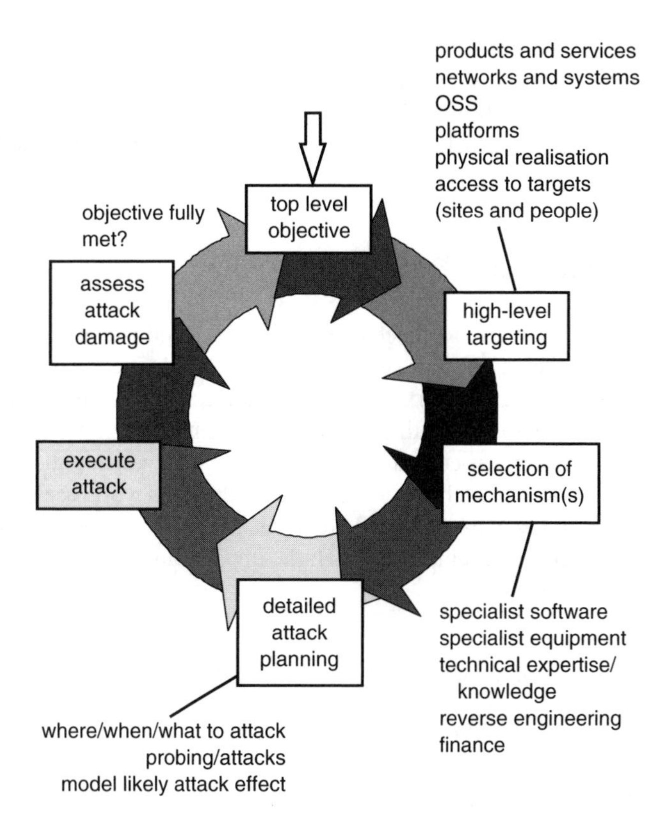

Fig 13.5 Planning cycle.

Table 13.1 provides examples of attack types and how they may fit into such categories, although not all of these are relative to this process.

Table 13.1 Sample attack types.

	Confidentiality	*Integrity*	*Availability*
Physical attack	Theft of documents, passwords, shoulder surfing, espionage, 'dumpster diving'	Changing the contents of documents	Damage to or destruction of equipment, sites, personnel.
Electronic attack	Hacking, copying electronic data, espionage	Data manipulation, disruption to config-uration settings or records	Virus, data deletion, power spiking, change passwords, damage to disaster recovery processes or facilities
Psychological attack	Social engineering	Perception management	Threat, blackmail

The outcome of the assessments should be a set of varying scores across the range of protagonists. This is done to provide better focus and priorities for clients who may have limited funds available to implement security countermeasures, i.e. which vulnerabilities and which protagonists present greatest concern and need to be addressed first?

13.9 Detecting and Reacting to Attack

Detecting a malicious electronic attack event is not easy and is compounded by the lack of historical evidence of how such attacks may be defined, e.g. using attack signatures or footprints. IP networks are under routine and repetitive 'attack' (e.g. pings and port scans) but most of this low-level tinkering is identified and rejected by firewalls and is irrelevant compared with the few high impact attacks that are known. Attacks on public switched networks within the telco world are not common and those that are known have tended to be in support of fraud or have been perpetrated by disgruntled employees. High impact attacks may exhibit the following characteristics:

- malicious intent;
- planning, structure and targeting;
- attempts to thwart recovery;
- complexity and diversity.

BT is developing a specific event-reaction process to address these issues.

13.10 Benefits to Customers

Security is becoming an increasingly important part of the specification for services and solutions required by customers in all sectors and the demand for Internet security assurances is increasing significantly. It is important to avoid misalignment of customer expectation and delivered capability, particularly where the service may have IA implications. On the one hand, the effort and investment BT makes in providing security in products and solutions may not be appreciated nor valued by the customer, and on the other, the customer may believe that they are protected against certain threats for which they have, in fact, no protection. Neither situation makes for good and sustainable business or long-term relationships.

Malicious electronic attacks against the commercial sector are likely to be associated with large enterprises where the gains to be enjoyed by the attacker are commensurate with the investment needed to develop the attack. As businesses move towards placing greater reliance on their communications and IT infrastructure, and transfer more responsibility on to outsourcing suppliers, the opportunities for companies like BT for value-add business become significant. However, the further BT moves up the value chain with its customers, the more customers will expect from BT and the further it has to fall if it fails to meet expectations. The threat calls for a critical assessment of all products, services and platforms to ensure that the part that they play in maintaining the security of BT's or its customers' commercial activities is fully understood and that the risks are correctly managed.

BT is a trusted supplier of IT and communications services to Government and private sectors. The company has a long and respected record for the safe handling of national and international communications services and enjoys the confidence of a wide customer base. The work of the IAP is increasingly relevant to customers who are considering procurement or outsourcing contracts, which are characterised by the following themes:

- high commercial value;
- high regard for corporate reputation;
- emerging evidence of threat;
- high levels of integration;
- high usage of open systems;
- awareness of contribution to the CNI.

13.11 Future Development

The concept of the CNI is developing through the activities of the NISCC and IAAC, and key private sector players are moving towards the development of community-of-interest forums to discuss common needs and practices. Through the activities of the IAP, BT is perceived as a trusted member of this community and is well placed to serve as a consultant and supplier.

The threat model and its associated processes are reviewed after each use with the aim of identifying improvements. Previous studies have been weighted towards a vulnerability-led perspective and this is now being complemented by a threat-led perspective, including the use of 'susceptibility to targeting analysis' techniques. This may highlight new factors for vulnerability assessments based on the planning and intelligence-gathering processes that are employed by organised attackers. BT is also assessing the results of protagonist model results over time to identify trends and opportunities for direct linkage with detection and reaction processes; these may be external to BT as well as internal.

BT's CERT services are also being actively marketed outside BT due to demand from customers for outsourcing Web security and the need to present a comprehensive security services portfolio (see also Chapter 12) [11].

References

1 Bennett, S. R.: '*Opening statement*', Joint Economic Committee, Cyber Threats to US Economy (February 2000).

2 Institute of Chartered Accountants in England & Wales: '*Internal Control — Guidance for Directors — The Combined Code*', (September 2000).

3 CSI/FBI: '*Issues and Trends*', Computer Crime and Security Survey (March 1999).

4 US Chamber of Commerce report (February 1999).

5 KPMG: '*Information Security Survey 2000*', (April 2000).

6 '*Joint Doctrine for Information Operations*', US Joint Chiefs of Staff (October 1998).

7 Forum of Incident Response and Security Teams — http://www.first.org/

8 IAAC Dependencies and Risk Working Group: '*Green Paper*', (July 2000).

9 Colwill, C. J.: '*Large Scale Risk Analysis*', IAAC Paper (July 2000).

10 Colwill, C. J.: '*IA Protagonist Model*', IAAC Paper (February 2001).

11 BT Ignite — http://www.btignitesolutions.com/solutions/security/

14

BIOMETRICS — REAL IDENTITIES FOR A VIRTUAL WORLD

M Rejman-Greene

14.1 Introduction

Electronic transactions and processes are irrevocably changing the lives of people throughout the world, whether it be at work or in the home. As with the introduction of the letter post and the telephone, increasing interaction with people and systems remotely — rather than face-to-face — brings new challenges to society. Using the Internet, there are even fewer clues as to the identity of the sender of a message, except for references in the content or context of the communication. Individuals can take advantage of this opportunity for anonymity, or use pseudonymous identities in novel ways that would be impossible without such features of a new medium. However, there are many services that require the recognition of the identities of specific individuals. It is for these services that biometric methods — secure automated methods of recognising individuals using a measurable, distinctive physical aspect or action — offer the prospect of more secure ways of authentication. Figure 14.1 lists a number of commercially available biometric methods and a selection of those that are currently under development.

There are relatively few ways, using a wholly electronic interface, of verifying the identities of people:

- proof of identity by demonstrating knowledge of a secret known only to the individual — examples of such knowledge are passwords, personal identification numbers (PINs) and the ability to recognise a face from a block of randomly selected facial images [1];
- proof by possession and use of a unique, unalterable token held by one person — among such tokens are smart cards with stored personal identifiers and USB tokens storing digital certificates;

- proof by geographical position, with evidence of a user being at a specific house or office location, e.g. caller identification is often used in systems that operate over the public telephone network;
- proof by successful confirmation of a distinctive feature of the individual's body — among these physiological biometric methods is AFR (automatic face recognition), the comparison of a person's face, as captured by a camera, with a previously validated representation;
- proof by a distinctive, measurable and stable action by that individual, one such action being the time-analysed electronic form of a written signature (DSV, dynamic signature verification) — such behavioural biometric systems operate, in general, under the control of the person and may be less robust than physiological systems.

physiological methods	**behavioural methods**
commercially available systems	
fingerprint patterns	dynamic signature verification
face recognition (and infra-red)	speaker verification
eye: iris scanning	keystroke dynamics
eye: retinal scanning	
hand geometry	
systems under development	
vein patterns on back of hand	gait recognition
body odour	
lip patterns and movements	

Fig 14.1 Examples of biometric methods.
(Note that most methods are not exclusively physiological or behavioural.)

Whatever approach is chosen, system designers must make assumptions about the link between a person and the result of the authentication challenge that the system presents to that person. Often these assumptions are not clearly articulated. For example, there is a significant body of research demonstrating that users do not keep passwords secret, writing them down 'in a safe place', or even sharing them when working on a group task [2] (see also Chapter 15). It is also widely acknowledged that tokens are shared. A holder of a card may lend it temporarily to a colleague who has mislaid their own card, recognising that waiting for a re-issue of the card could delay critical work. Indeed, the helpfulness of people is a continuing challenge to the security professional. When systems fail to operate as intended —

or are made to fail intentionally — back-up operations can offer an opportunity for impersonation by an impostor. The proponents of the use of biometric methods (the fourth and fifth of the methods listed above) claim that fewer assumptions are needed for these approaches, since these methods authenticate the person directly — not a proxy for that identity.

The security requirements of a service may not be met by one authentication method alone. Two or more methods could be required, either taken together at the start of a session or invoking a further request at a critical part of a process, such as authorisation of payment for an electronic transaction. Care is required in interpreting the added security value that such a combination of methods may offer.

14.2 Using Biometric Systems

A biometric system consists of the following four components:

- a device or sensor that measures the characteristic action or feature;
- an algorithm that processes the signal and compares it with a standardised representation of the individual's biometric feature or action;
- a decision module that determines whether the comparison is acceptable and passes this result to the application;
- a management framework and supporting processes.

Conventionally, biometric systems are designed to operate in one of three modes.

- Identity verification

  Most common applications verify the identity of users. In this mode, subjects make a claim to being a specific individual, either directly by typing their user name, or offering an identifier such as a uniquely allocated number. This identifier points to their template, a record in a database obtained at a previous enrolment and summarising the essential features of the biometric dataset for that individual. For example, in a DSV system, the template may consist (in part) of a set of measurements representing the time intervals between the pen being raised and then replaced on to a digitising tablet when a user signs her name. The format of the template is designed to facilitate comparison with measurements made when the user's identity needs to be verified. Alternatively, a template could be stored locally for a user — for example on a smartcard, and the claim to a specific identity made by the act of inserting the card into a reader.

- User identification

  The biometric system captures the data set of the single user (e.g. an image of a fingerprint), processes it to remove noise and extracts key determining features for comparison with the entire database of templates. The template that corresponds most closely to the feature set of the subject is deemed to identify the subject. In this 'walk-up' mode, the user makes no claim of identity, minimising authentication overheads for the individual.

- Multiple identity screening

  As there are many applications (e.g. social security payments by governments and the issue of driving licences) where criminals can gain a benefit fraudulently by generating multiple identities, it is essential to identify that a user is not already on a database. In these applications, the search algorithm uses the applicant's dataset as collected by a sensor in the biometric device to make a comparison with all of the stored templates. If no templates offer a close match, the user is permitted to enrol on to the database as a unique, fresh individual.

One common factor in all these types of application is the comparison of a measurement taken at the time of authentication of the individual with a measurement made previously at the time of enrolment. Although this matching may appear to be a trivial step, the variability of conditions at the human/machine interface results in considerable problems in the practical implementation of biometric methods. Users may position their fingers at a slightly different angle in a fingerprint biometric, the ambient lighting in an AFR system may change from morning to evening, a user attempting to verify by means of a speaker verification scheme may develop a cold, etc. A biometric system that is configured to demand too close a match can reject the correct user, causing them to resort to a back-up system, which itself may be less secure. Conversely, if the threshold for accepting a valid match is set too low, the system may allow an impostor to be accepted (see Fig 14.2).

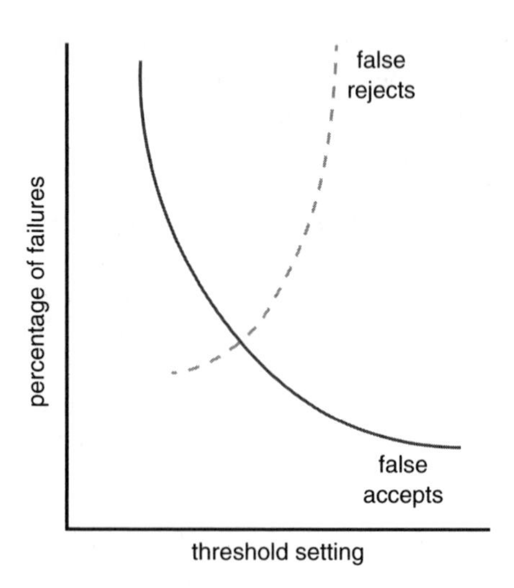

Fig 14.2 Adjustments of the security setting.

The trade-off between false rejection and false acceptance determines the usability and security of a system that employs a biometric. In many systems, the administrator can adjust the threshold value, even individually for each user, setting this balance in line with the system security policy. An alternative approach to making a yes/no decision assesses the probability that the person is who they claim to be; this requires modelling of the distribution of the biometric features among the population of all users. This probability can be input into a more complex decision-making programme that takes account of other relevant information, e.g. the context of the transaction.

Clearly, it would be useful if these performance characteristics were known in advance of decisions about which biometric to use. It would guide users in the selection of optimum threshold settings corresponding to a specific security policy. Experience shows, however, that these values are very sensitive to the context of the application, the demographic details of the subject population (gender, age, disability, race, etc), and the willingness of participants to co-operate, etc. The results of tests on systems with biometric modules are particularly controversial and, as will be shown later, it is only recently that ground rules for testing have been agreed.

The decision to deploy a biometric as part of a service requires careful assessment of the whole life costs of their introduction and use. Early applications, such as the Expo92 fingerprint-enabled season ticket, demonstrated the unexpectedly high costs of enrolment, with users requiring significantly more support to become familiar with the new technology. Results of laboratory tests may give an over-optimistic assessment of the performance of devices in the field, as evidenced by the initial experience of immigration authorities introducing fast-track procedures through international airports using hand geometry devices. Further development work may be needed to adapt systems to such novel situations. Additional costs may also be incurred if templates are to be updated securely as users grow older. It should be noted, however, that the existing methods which a biometric system is to replace may have surprisingly high ongoing costs — a major organisation may need to maintain a large call centre solely to reset forgotten passwords.

Biometric methods can contribute to the security of systems in other ways. The act of submitting a fingerprint, or signing an electronic document using DSV methods, can increase the perception of accountability for a user's actions. Excuses of delegation of responsibility or accidental completion of a transaction will be less convincing. For example, DSV biometrics have been used successfully in signing off test results on pharmaceutical products, so as to meet the requirements of the US Food and Drug Administration's medical testing protocols. There have also been recent designs for responsible anonymity using attribute certificates in public key infrastructures. A secure, supervised enrolment to a fingerprint sensor collocated on a smartcard, with tamper-resistant template and certificate storage, could be a component of such designs; however, such complex hardware is still under development.

14.3 Performance and Testing Issues

An ideal biometric method would exhibit characteristics such as:

- uniqueness — a different measure for every person on the planet, with sufficient differentiation to offer secure operation over a range of template-matching thresholds;
- ease of access to that measure — either because it is visible in situations where it is needed, or it is an action that is performed routinely in that context, and yet difficult to access if the person does not want to be identified;
- ease of electronic capture of that measure using cost-effective, robust and compact digital sensors;
- stability of that measure from birth to death;
- fast operation, in line with the expectations of users in a specified application;
- fast and easily coded algorithms to extract the relevant features — to filter out unwanted components and compare with features coded in the template;
- being unaffected by environmental conditions;
- a good match to the application that is to be secured;
- safety and security, with the biometric template being mathematically hard to deduce and copy;
- acceptability both to the user community and to the courts in the event of legal disputes.

No biometric method or system satisfies all of these requirements, and the deployment of services that will make use of a biometric must make allowance for those characteristics that fall below the optimum. For example, no biometric can guarantee uniqueness. Iris recognition systems, with effective template sizes of 256 bytes, have the potential of distinguishing between members of large populations, based upon research data obtained from thousands of pairs of eyes [3]. In contrast, hand geometry methods offer much less potential for this and are deployed in systems where verification, rather than identification, is required. The template size of 9 bytes in one commercial implementation has other advantages, allowing compact storage, for example, on magnetic stripe cards. Although one hundred years of forensic policing has testified to the practical uniqueness of fingerprints, the requirements of price, automation and decision time result in commercial biometric packages that cannot fully exploit this potential.

Some physiological biometric identifiers appear to exhibit lifetime stability. Fingerprints, and probably iris patterns, offer this long-term robustness. However, changes in the body with age can bring problems with skin elasticity and arthritis making the use of hand geometry systems more difficult. Other biometric implementations may need to offer template-ageing programmes, by periodically adjusting the parameters in users' templates.

Limitations of measurement access to the biometric feature may determine the type of biometric that can be used. For example, in many laboratories and semiconductor clean rooms, employees need to wear protective gloves, which precludes most commercial fingerprint systems, but which could still allow the use of hand geometry devices. Of the two distinct types of eye biometric, retinal scanning — the older system, in which the pattern of blood vessels behind the ball of the eye is registered — requires active co-operation of the subject to look into a custom optical system. In contrast, the iris of the eye (the coloured element) is visible to all, albeit currently still requiring a level of co-operation on the part of the user.

Over the past five years, sensing devices have become substantially smaller and cheaper offering the prospect of increased application of biometrics in the corporate electronic security market. Small CMOS cameras are beginning to be integrated into laptops and the size of fingerprint sensors has shrunk one hundred-fold with the advent of single-chip matrix sensors consisting of a postage-stamp-size piece of patterned silicon.

Semiconductor devices are potentially cheaper to fabricate in large quantities and can be integrated into more complex hardware designs offering data storage, algorithm processing and encryption engines. The $25 fingerprint sensor is now a commercial prospect.

Which methods work best? It is easy to forget that one biometric method can be implemented in many different ways. The algorithm that is selected for processing the image into a template will influence the speed of operation and accuracy in use, although there have been few comparative studies of their efficiency in this respect. (One exception is the continuing series of FERET comparisons of the accuracy of algorithms for facial recognition [4].) For example, there are over 30 suppliers of fingerprint biometric systems, some using common hardware, but offering different ways of processing the signal from the sensor. Most use a similar approach to that of modern police AFIS (automated fingerprint identification system) schemes, searching out minutiae — characteristic features on the scale of the fine ridges that make up a fingerprint. These features include abrupt ends of ridges and places where a ridge splits into two. From this pattern of characteristic features, the template can be coded in many ways, with some systems storing the relative positions while others record the number of ridges between pairs of minutiae points. One supplier uses the complete fingerprint image and codes the moiré pattern resulting from an offset overlap. Still others input the images or data points directly into a neural network.

The problem of testing of devices and systems has been mentioned earlier. Early tests were performed by manufacturers themselves on very limited numbers of subjects and under idealised laboratory conditions. Performance measures were extrapolated from small samples or using unproven models, with many suppliers quoting unrealistic figures. A decade ago, in an attempt to standardise on common procedures, Sandia National Laboratories in the USA conducted two series of comparative tests on stand-alone biometric devices and included questionnaires to elicit impressions of acceptability from participants [5]. These tests demonstrated the need for further research to understand the complexities of the human/machine interface, the effect of the environment, etc.

A number of initiatives have transformed our understanding of testing of biometric devices and systems — studies at the biometrics research centre at San José State University [6], the encouragement of the Biometrics Consortium [7] (a focal point for the US Government's research, test and development activities), the European Commission's BIOTEST project [8] led by the UK's National Physical Laboratory, and the formation of the UK Government's Biometric Working Group [9]. The last group has published a guidance document of best practices in the testing of biometric devices [10] and recently comparative tests on a range of commercial devices have been completed using this standardised procedure [11]. Examples of face, fingerprint, hand geometry, iris scanning, hand-vein pattern and speaker verification biometric systems were examined in an environment typical of many offices. Prospective designers should not use the results of these studies to identify 'best buys'. System performance depends critically on both the application and the operational environment. In this particular application, the users were, in general, technically aware, the ambient lighting levels were controlled, and voice verification systems were sited in a quiet area. It has been long recognised that facial recognition systems are especially susceptible to variations in the angle of incident light and a trial of a banking automatic teller machine using speaker recognition was abandoned once ambient noise levels were found to interfere with the operation of the algorithm. Among the interesting outputs of the comparative study was the fact that the devices appeared to work best for younger subjects and that, for the most part, men performed better than women, perhaps reflecting the profile of the developers of such systems.

14.4 Security of Biometric Devices and Systems

A biometric cannot be treated as a simple 'drop-in' replacement for existing authentication systems. The complex processing, specialised hardware and unproven performance of many systems offer numerous challenges to the designer of secure systems [12]. As part of the process of defining security requirements for operation of biometric devices in systems, the Biometric Working Group has commissioned a 'protection profile' for biometric devices [13].

Some vulnerabilities are easy to predict. In the case of facial recognition systems, the incidence of identical twins in the population places a limit on most systems (except, perhaps, for those that use the spatial variations in heat emitted from sub-surface blood vessels in the face). Nevertheless, the most direct attacks on systems can be the most effective — cameras can be physically disabled by obscuring the front window with paint or objects placed in front of them. Other vulnerabilities are more subtle. Many suppliers of devices claim that 'live-and-well' features are implemented in order to counter attacks by simulation; however, a recent paper casts doubt on these claims, with the authors reporting that they were able to enrol plastic dummy fingers with stamped fingerprint patterns on many commercial units [14]. As in many complex systems, design flaws can offer unintentional 'back doors' to critical parts of the unit. Such flaws may lie undiscovered, often due to the unwillingness of suppliers to publish details of their implementation. In one system recently tested at Adastral Park, an artefact of the algorithm allowed an unusual template (obtained during a normal enrolment) to match successfully against 15% of the rest of the trial group.

Finally, no system designer should forget the weakest link in a security system — the people working inside an organisation. For example, unless there are proper security policies to which users adhere, it may be relatively easy for unauthorised users to collude with managers at an enrolment session to add multiple instances of their identity or to link their biometric template to a false identity.

14.5 Legal and Acceptability Issues

A biometric template or related data can be traced back to an individual [15], and therefore, in the European Union, it is likely to be subject to the provisions of the 1995 European Directive on Personal Data Protection [16] and the laws of individual member states which incorporate this directive into national law [17]. The major provisions of such laws are summarised in eight principles, among which are provision of appropriate security measures for the databases and systems storing the data, fair dealing in the collection and use of the data and limitations on transfer or exchange of data outside the European Economic Area. One recommendation is that applications should be segmented by sector or division of an organisation so that, where the application does not require it, large databases of data are not collated [18]. The importance of this concern beyond the EU has been recognised by the International Biometric Industry Association (IBIA), an industry lobbying group headquartered in Washington, which has agreed a set of privacy principles, advocating clear policies for applications in the private sector and enforceable legal standards for government deployments [19].

In the commercial environment, consumer acceptability of technology can never be taken for granted. There are conflicting views on whether biometric systems will be accepted in wide-scale deployments in Europe. Small-scale testing by one consumer group showed that users looked favourably on the use of fingerprint biometrics to secure access to their bank accounts. Other evidence points to concern with the association between criminal activities and fingerprints. Still others are concerned that the biometric image, rather than the template, may be stored — and then misused. Once compromised, the biometric approach of that supplier's system may be unusable, and consumer trust in biometric systems will be diminished. Other concerns of users centre on the increased accountability for actions — no excuses anymore, no delegation of authority without responsibility. Managers may not worry about this, but if these views are held by significant minorities in the workforce, biometric systems may not operate as successfully as predicted.

Although most studies into user acceptability have been conducted through questionnaires, there is a developing view that more sophisticated methods are required to elicit deeply held concerns. Among these techniques are qualitative approaches such as grounded theory methods [20] that attempt to limit the impact of the preconceived views of the investigator.

14.6 Standardisation Activities — Evidence of a Maturing Technology

Many of the suppliers in the biometrics market-place were small companies, often underestimating the need for investment in associated activities such as testing and security. Many are no longer in existence. This crisis of confidence in the future of the industry was a driver in the development of standards for interoperability. The wide variety of biometric methods, devices, algorithms and applications led to differing views on the best standards architecture. A number of separate application-level initiatives were brought together in the BioAPI Consortium, which released an updated version of the BioAPI specification in March 2001, along with a reference implementation [21]. A device-level standard was judged to be too complex, although in May 2000 Microsoft and I/O Software announced their co-operation in the development of BAPI, an alternative biometrics API, with the aim of integrating biometric authentication into future versions of the Windows operating system [22].

More biometric- and industry-sector-specific standards are envisaged. Already, the financial sector has agreed a standard for biometric management and security [23], and standardised formats for the exchange of biometric data files are also available now [24].

14.7 The Future of Biometric Methods of Authentication

For over two decades, the supporters of biometric technologies have predicted that the market would explode and that volume consumer applications would render obsolete the password and PIN. Clearly this has not happened. Nevertheless, over that period, devices and algorithms have decreased significantly in both size and cost. The preferred biometric method has also changed. Whereas 10-20 years ago, speaker verification and signature verification were regarded as contenders for early adoption, now it is the fingerprint biometric that offers the most interesting mix of price, performance and robustness, even though there are still some concerns about their susceptibility to electrostatic discharges. Two years ago, Compaq introduced the first modern consumer biometric device — a fingerprint-enabled computer mouse priced at under $100. This was followed by the availability of computer keyboards with fingerprint sensors. Both used optical sensors that were too large for the growing number of portable terminals that arguably need this form of security more than desktop machines. The availability of compact silicon IC devices is likely to stimulate the application of biometrics in laptops, PDAs and mobile telephones, provided that there is a business case for their widespread introduction (see Fig 14.3). Furthermore, if public key infrastructure (PKI) deployments increase, it is anticipated that the demand for a more secure method of activating the private key will expand. Indeed, some suppliers have partial solutions available now. A future vision of a single smartcard for proving personal identity is no longer unrealistic. It is likely to feature an on-board sensor, the algorithms for processing images from the sensor and for comparison with a securely stored template, together with a protected digital certificate and private key.

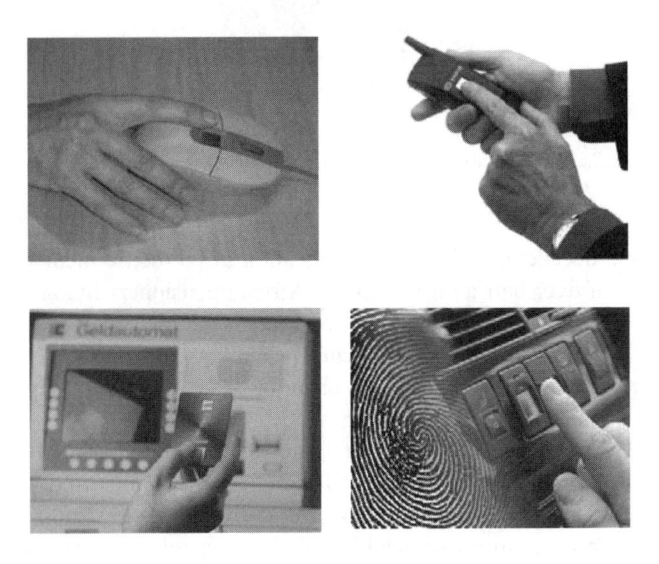

Fig 14.3 A world of fingerprint terminals.
[Photographs: Infineon, Sagem]

Small iris-scanning devices are also available now (see Fig 14.4), and, for applications requiring higher performance, these may be an alternative choice to fingerprint biometrics. Of other biometric methods, hand-geometry devices continue to sell well as an alternative to door locks in buildings needing high physical security, although fingerprint-enabled systems have a place in this sector as well. Fingerprint systems are also making inroads into the time-and-attendance market, ensuring that only authorised employees are on a company's premises. Speaker verification and facial recognition approaches are also being deployed, and there is some interest in using more than one biometric to improve the performance and reliability of authentication solutions. New DSV signature systems continue to appear, with special pens now available that dispense with digitising tablets by including all of the sensing and processing components in the body of the pen.

Fig 14.4 Iris scanning desktop PC peripheral. [Photograph: Iridian]

More than a decade ago, the major application of biometric methods was in the authentication of over half a million South African pensioners living far from post offices and cash dispensers. Since then, much of the momentum for their introduction has come from governments, whether it be in social security and driving licence applications in the USA or in the issue of electoral registration and citizen's cards in newly developing countries.

For the private sector, the future of large-scale deployment and use of these technologies is more difficult to discern. The increasing number of trials testifies to the continued interest in the possibilities offered by biometrics, and the number of deployments continues to grow steadily, albeit using relatively small numbers of devices. When — or, if ever — every electronic terminal will have biometric authentication remains an open question, but the sound technical understanding gained over the past decade and the maturity of the systems offered by suppliers can only increase the likelihood that passwords and tokens will be replaced in the working lifetime of most readers.

References

1 ID Arts — http://www.idarts.com/

2 Adams, A. and Sasse, M. A.: '*Users are not the Enemy*', Comms of the ACM, **42**(12), pp 41-46 (December 1999).

3 Gifford, M. M. et al: '*Networked biometric systems — requirements based upon iris recognition systems*', BT Technol J, **17**(2), pp 163-169 (April 1999).

4 US Department of Defense — http://www.dodcounterdrug.com/facialrecognition/FRVT2000/frvt2000.htm

5 Holmes, J. et al: '*A performance evaluation of biometric identification devices*', Sandia National Laboratories report, SAND91-0276 (June 1991).

6 San José State University — http://www.engr.sjsu.edu/biometrics/

7 The Biometric Consortium — http://www.biometrics.org/

8 National Physical Laboratory — http://www.npl.co.uk/npl/

9 Communications-Electronics Security Group — http://www.cesg.gov.uk/biometrics/

10 CESG — http://www.cesg.gov.uk/biometrics/pdfs/Best%20Practice.pdf (Version 1.0, January 2000)

11 CESG — http://www.cesg.gov.uk/biometrics/pdfs/Biometric%20Test%20Report%20pt1.pdf (Issue 1.0, March 2001)

12 Rejman-Greene, M.: '*Security considerations in the use of biometric devices*', Information Security Technical Report, **3**(1), pp 77-80 (1998)

13 CESG — http://www.cesg.gov.uk/biometrics/pdfs/Biompp081.pdf (Draft Issue 0.81, January 2001)

14 van der Putte, T. and Keuning, J.: '*Biometric finger recognition: don't get your fingers burned*', Smart Card Research and Applications, IFIP Fourth Working Conference, pp 289-303 (September 2000).

15 Grijpink, J.: '*Privacy Law: Biometrics and Privacy*', Computer Law and Security Report, **17**(3), pp 154-160 (2001).

16 Directive 95/46/EC of the European Parliament and of the Council: '*On the protection of individuals with regard to the processing of personal data and on the free movement of such data*', (October 1995) — http://europa.eu.int/eur-lex/en/lif/dat/1995/en_395L0046.html

17 In the UK, the Data Protection Act (1998).

18 Davies, S.: '*Touching Big Brother: How biometric technology will fuse flesh and machine*', Information Technology and People, **7**(4) (1994) — http://www.privacy.org/pi/reports/biometric.html

19 International Biometric Industry Association — http://www.ibia.org/privacy.htm

20 Glaser, B. G. and Strauss, A. L.: '*Discovery of grounded theory*', Strategies for Qualitative Research, Aldine (Chicago) (1967).

21 BioAPI Consortium — http://www.bioapi.org/

22 I/O Software Inc — http://www.iosoftware.com/press/

23 American Bankers Association, X9 Committee (ANSI X9.84 Biometric Management and Security Standard) (2001) — http://www.x9.org/

24 National Institute of Standards and Technology (Common Biometric Exchange File Format) (Jan 2001) — http://www.itl.nist.gov/div895/isis/cbeff/

15

TRANSFORMING THE 'WEAKEST LINK' — A HUMAN-COMPUTER INTERACTION APPROACH TO USABLE AND EFFECTIVE SECURITY

M A Sasse, S Brostoff and D Weirich

15.1 Introduction

With the exponential growth of networked systems and applications such as eCommerce, the demand for effective computer security is increasing. At the same time, the number and seriousness of security problems reported over the past couple of years indicates that organisations are more vulnerable than ever. In many of the reported cases, user behaviour enabled or facilitated the security breach. The security research community — which hitherto largely ignored the human factor — now acknowledges that:

> '... security is only as good as it's weakest link, and people are the weakest link in the chain.' [1]

The opposition recognised and exploited this state of affairs earlier. Kevin Mitnick, arguably the world's most famous hacker, testified to the US Senate committee that he had obtained more passwords by tricking users than by cracking. In his new role as security evangelist, he never ceases to point out that:

> '... the human side of computer security is easily exploited and constantly overlooked. Companies spend millions of dollars on firewalls, encryption and secure access devices, and it's money wasted, because none of these measures address the weakest link in the security chain.' [2]

The first implication of this new perspective on security is that the traditional security approach — addressing the problem by developing ever more complex technology — is not sufficient. We agree with this conclusion. However, labelling users as the 'weakest link' implies that they are to blame. In our view, this is a repeat of the 'human error' mindset that blighted the development of safety-critical systems until the late eighties [3]. Consider the following examples of users violating password rules.

- Ambushing

  A user is told that his password has expired, and he must change it immediately or be locked out of the system — he feels stumped, and ends up choosing his wife's name. This is exploited by a colleague who wants to look at files he has no permission to access. He tries out the user's family members' names to get into the system, and succeeds. Many password systems 'ambush' users without warning. People have difficulty designing and memorising strong passwords, and they have even more difficulty doing so when under pressure.

- Conflicting goals

  An aircraft designer needs to access six different systems; company policy states she must have a different password for each. The system only accepts strong passwords, and requires a change each month. Recently, she was reprimanded by her boss for missing an important deadline — she worked on a Sunday but could not get to some files because she could not recall the right password and could not get help. She now keeps a note with her current passwords under her mouse mat, where an industrial spy working as a contract cleaner finds it and uses the passwords to download confidential design drawings. The security mechanism and policy created an impossible memory task (at least without instructions or training). When failure at the memory task interfered with her work, the organisation failed to recognise and address it. The user was left with two conflicting goals and forced to relegate security to second place.

- Requested disclosure

  A hacker calls company employees and tells them that he works for IT support and needs their password to update some programs on their machine. Since no admin accounts have been set up on many PCs in this company, IT support staff often need to ask users for their passwords when they want to get into these machines. This is a contextual issue — if systems are set up so users are regularly asked to disclose their passwords, it is difficult for them to distinguish in which context disclosure is safe, and when it is not.

These examples illustrate issues that are often overlooked in security design. As Adams and Sasse [4] pointed out, security has largely ignored usability issues; many users of security systems face unattainable or conflicting demands, and receive no support or training. A human-computer interaction (HCI) design approach takes into account that users and technology work together completing a task (in order to achieve a goal) in a physical and social context. Any specific user, technology, task or context may bring constraints to the design problem. In the remainder of the chapter, we systematically examine the following issues involved in security design, and outline how HCI knowledge can be employed to address them:

- technology;
- user;
- goals and tasks;
- context.

While these issues can affect any security mechanism, we are using examples from our research on one specific mechanism — passwords — to illustrate them. Our recommendations are based on an analysis of the relevant literature, and empirical findings from the following four studies.

- Study 1

  By use of questionnaires, 144 BT employees (half at management grade or above) were asked both to describe the cause of the last password problem they encountered that led to a password reset, and also to report their number of overall passwords at work and how they used them.

- Study 2

  This study analysed six months of password reset logs from the BT password helpdesk.

- Study 3

  In-depth interviews on attitudes to passwords were conducted with 17 users (10 were BT employees, 6 were PhD students at UCL, and one a systems administrator working in finance). The interviews lasted between 30 and 60 minutes and were subsequently transcribed for analysis.

- Study 4

  32 students used a Web-based system to practise on and also to submit assessed coursework [5]. System logs of the passwords and passfaces (see section 15.3.1) used allowed us to study not only frequency of logins and login failures, but also what caused the wrong selection.

15.2 Technology

There are five ways to authenticate users (as detailed in Chapter 14) most security mechanisms use a two-step procedure in which identification and authentication are combined. An example of such a combination are cash cards (token-based identification) combined with a PIN (knowledge-based authentication). By far the most common access control mechanism in computing is the combination of a userid (identification) and password (authentication). Most password systems are implemented in the same way, as described below.

- Userid and password

  The mechanism issues a userid for every new user, and also a password (which can be changed by the user to one of their choice). The password is supposed to be a secret shared between the user and the computer only, it should not be disclosed or written down.

- Log-on

  To log on, the user has to enter their userid and password, which the system processes and compares to the entries it has stored. If it finds a match, the user will be given access to the computer system, but if there is no match, the user will not be allowed access, and may have to contact a system administrator to have a new password issued. Many systems suspend an account after 3 or 5 unsuccessful login attempts, and bar further attempts until the account has been re-set.

From a technical point of view, the password mechanism is a low-cost option, the risks of which are well understood. It is also a mechanism with which many users are familiar. However, there are also a number of usability issues connected to its use.

- Number of passwords

  Password mechanisms are usually implemented on a per-system basis. This means that users need to log into each password-protected system individually; the time required to log into a number of systems several times a day can add up. Some operating systems store userid and password and automatically use it on the user's behalf (e.g. to mount remote volumes).

- Password policies

  The growing number of systems with which users have to interact creates memory problems (see section 15.3.1). The problem is often exacerbated by password policies, usually based on the Federal Information Processing Guidelines [6]. These rules state, for instance, that:

  — passwords must be strong, i.e. a pseudo-random mixture of letters, numbers and characters;

  — users should have a different password for each system;

— passwords should be changed at regular intervals, and accounts of users who do not comply deleted or suspended.

- Varying systems

  There is great variability of userids and passwords across different systems, e.g. Unix takes up to 8 characters, Windows 95/98 up to 14, and Windows 2000 up to 127. Some systems have highly elaborate content restrictions (e.g. "there must be at least three non-letter characters in the password, and letter 4, 5, or 6 must be such a character"), but these vary from system to system.

The result is a huge demand on users' memory:

- users not only have to remember passwords, but also the system and userid with which it is associated;
- users have to remember which password restrictions apply to which system;
- users have to remember whether they have changed a password on a particular system, and what they have changed it to.

It is thus not difficult to accept that many users cannot cope. As a result, the cost of re-setting passwords has reached significant levels in BT and elsewhere. In response to rising cost and user protests, many organisations give users permission — or even direct them — to write their passwords down and keep them in a safe place. This violates the first principle of knowledge-based authentication — that the password should exist only in two places — in the system (in encrypted form) and the users' mind.

A technical solution to reduce the number of passwords is a single sign-on (SSO) login system, which many companies are starting to deploy. This reduces not only users' memory load, but also the total amounts of time users spend on logins. If SSO is not feasible (e.g. because of cost), allocating users a single userid for all systems, standardising password rules, and enforcing them consistently, can improve the situation somewhat. Most security policies decree that users should have different passwords for different systems, to limit the number of systems compromised if an unauthorised person gets hold of a password. From a usability point of view, allowing users to have the same password on different systems is desirable because it increases frequency of use and memorability (see section 15.3). Ultimately, this makes for more effective security because it gives users a chance to have strong passwords they can remember, and a strong password reduces the chances of it being compromised in the first place.

15.3 User

15.3.1 Memorability Issues

The user characteristic that has the primary impact on password design is memorability. There is a huge body of research on human memory, but the most important issues related to passwords can be summarised as follows:

- the capacity of working memory is limited;
- memory decays over time — this means people may not recall an item, or not recall it 100% correctly;
- recognition of a familiar item is easier than unaided recall;
- frequently recalled items are easier to remember than infrequently used ones, and retrieval of very frequently recalled items becomes 'automatic';
- people cannot 'forget on demand' — items will linger in memory even when they are no longer needed;
- items that are meaningful (such as words) are easier to recall than non-meaningful ones (sequences of letters and numbers that have no particular meaning);
- distinct items can be associated with each other to facilitate recall — however, similar items compete against each other on recall.

Knowledge-based authentication mechanisms, such as passwords, require users to memorise items and recall them when accessing a specific system. Asking users to recall a single password and userid for one system may seem reasonable, but with the proliferation of passwords, users are increasingly unable to cope.

15.3.1.1 Password Problems

Before study 2 (the analysis of password resets), BT security staff believed that the rising number of password resets was due to a small number of careless 'repeat offenders' — by their own definition, employees who ask for a reset 6 or more times a month. Study 2 found that 91.7% of resets were caused by 'normal users', i.e. more than 90% of users cannot cope with the password mechanism in the way that was expected of them, which is a rather damning result in terms of usability of password mechanisms. The study also found that 30% of the 'repeat offender' helpdesk calls could be traced back to temporary staff, and were largely due to administrative (rather than memory) problems.

Study 1 (BT employee questionaires) is the first systematic study of passwords in a population of real users during their normal work. The average number of passwords per user was 16. We asked users to describe the cause of their last password problem, as well as the frequency with which they used the password. The responses were then categorised according to these frequencies into three groups — light use (from once per year to just under once a month), medium use (from once a month to once a day) and heavy use (used more frequently than once a day). The results are shown in Fig 15.1. Our data supports the findings from the Web survey by Adams et al [7], in which users reported that infrequently used passwords are the ones that are most often forgotten.

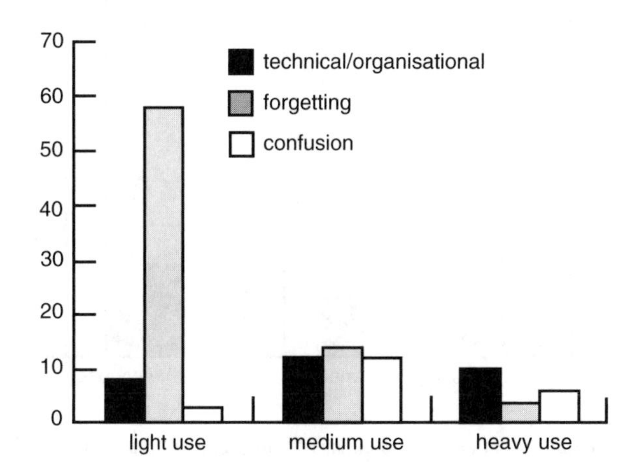

Fig 15.1 Frequency and cause of problems with passwords.

Study 1 also demonstrated the effect of password content. Figure 15.2 illustrates problems with a voicemail system using a six-digit PIN. The proportion of reported incidences where this PIN could not be remembered was very high — even when it was moderately or heavily used. While frequently used computer passwords can be recalled after periods of non-use, even heavily used PINs are forgotten after short periods of non-use.

In the discussion to date, login failure has been usually described as 'users forgetting passwords'. Studies 3 and 4 found, however, that users hardly ever drew a complete blank — the login usually failed because:

- they recalled the password partly, but not 100% correctly;
- they recalled a different password from the one required, i.e. a previously used password for the same system, or a password for a different system.

This shows the basic memory mechanisms (described above in section 15.3.1) in action — items decay in memory unless they are frequently recalled, and recall of similar items causes interference. The likelihood of 100% correct recall of infrequently used items is extremely low. This means that a password mechanism that demands 100% accurate recall every time is an extremely bad match for infrequently used systems. That the results for 6-digit PINs are even worse confirms the importance of password content. It also indicates that a token-PIN combination, often touted as a more usable replacement for passwords, is likely to cause more problems with infrequently used systems than a standard password.

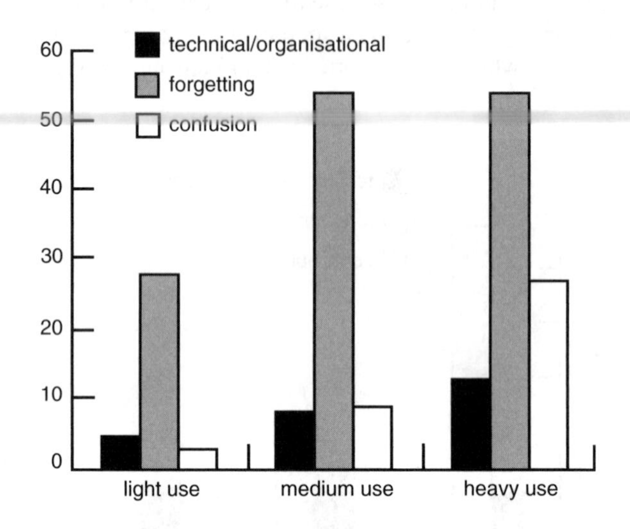

Fig 15.2 Frequency and causes of problems with a 6-digit PIN.

Heavy- or medium-use passwords were more often confused than lightly used passwords (see Fig 15.1). Heavily used passwords were more often confused than they were forgotten, and PINs are more frequently confused with each other than passwords (see Fig 15.2). Frequency of execution is one of the key considerations when matching technology to users' tasks (see section 15.4), but has to date rarely been taken into account in security. Our findings provide a powerful argument that existing authentication mechanisms are a bad match for systems that are not accessed on a daily basis, and that this causes the vast majority of password problems.

The results from studies 1 and 2 suggest that the other major cause of password resets are forced password changes. 13% of all reported password problems occurred just after changing a PIN. Moreover, the more frequently the PIN was used, the more problems occur after changing it. Approximately the same number of problems occurred just after changing moderately and heavily used passwords, whereas there were fewer problems with infrequently used passwords. There has been no systematic investigation of the impact of forced password changes on numbers of resets and password strength. The studies reported in this chapter provide some evidence that they make a significant contribution to password problems. The qualitative data in study 3 suggests that password strength declines with frequency of forced changes — as users become increasingly desperate in their quest for a password or PIN that stands out, their choices become more guessable.

*15.3.1.2 Can Passwords be Strong **and** Memorable?*

Both security [1] and usability experts [8] have stated that recalling strong passwords is a humanly impossible task because strong passwords are non-meaningful items and hence inherently difficult to remember. However, what makes a password easy to guess or crack is the fact that it is meaningful to many other people, as well as the password owner. It is possible to create passwords that are strong and meaningful — pseudo-random combinations of letters, numbers and characters that are meaningless to anyone but the password owner/creator. Many systems administrators use system-generated passwords for maximum security, and generate a sentence that describes, for instance, their opinion of a particular character in Star Trek (e.g. m,1aNib7 becomes 'Me, I am NOT impressed by SevenofNine'). They can manage a reasonable number of passwords by having different themes for different systems. For heavily used passwords, these pass algorithms seem to work well [9]. Zviran and Haga [10] obtained good recall results using a similar method with ordinary users, by giving them instructions to concatenate several words and interspersing them with characters for user-generated passwords (e.g. BF$gat0 for Black Forest Gateau). As part of our own research, we have incorporated a related set of rules into an on-line password tutorial. We have not completed an objective performance test (in terms of password strength and recall rates), but have received feedback that these rules helped users with password construction ('finally, I have a way of generating passwords the system will accept!') and recall.

15.3.1.3 Alternative Knowledge-Based Authentication Systems

One of the most fundamental HCI principles is to avoid unaided recall wherever possible, since it is known to place a considerable burden on users' cognitive load and overall ability to perform. There are authentication mechanisms that use cued recall and recognition, for example:

- composite weak authentication — many banks ask their customers to identify themselves by providing several weak but memorable items — this method may be used in combination with a password, and/or as a back-up method if the password has been forgotten;
- cognitive passwords [11] involve a series of questions about the user's personal preferences and history — after a certain number of correct answers, the user is considered to have passed authentication;
- associative passwords [10] employ word pair or phrase associations in a similar manner (e.g. Dear-God, Spring-Step, Black-White) while avoiding word association stereotypes — Ellison et al [12] refer to systems such as these that use the contents of episodic memory as employing personal entropy;

- the pass sentence mechanism [13] is an unaided recall mechanism in the first place — however, if the user does not get the sentence 100% correct, the user is prompted with questions about the pass sentence, and when the user answers enough questions correctly, login is allowed.

Systems based on recognition of visual items for authentication have received much attention recently:

- v-GO® [14] presents users with a visual scene — to authenticate, the user clicks on certain objects in the scene in particular order, and is then able to create a story using the objects in the scene, a mnemonic technique which aids the user in recalling the correct sequence of objects (v-GO® appears to have had only cursory analytic evaluation);
- both Deja Vu [15] and Passfaces™ [16] present users with panels of images, from which they have to recognise and select their pass images, the most significant difference between the two systems being the content of the images — Deja Vu employs randomly generated art, while Passfaces uses photographs of strangers' faces in an attempt to exploit people's ability to process and remember faces.

These systems have performed well in laboratory-style tests, producing recall rates of up to 80% even after up to 3 months of non-use [17, 18]. However, these results must be treated with caution because users in these trials were given one set of cues — this means there are no similar cues with which to confuse them. Unpublished trials with multiple cues indicate that problems similar to those with passwords occur when users are given several similar visual clues, or have to change them. In addition, the graphics involved in these systems require significant bandwidth and processing resources; when these are not present, login time can increase significantly, which in turn interferes with users' tasks (see section 15.4).

15.3.2 User Knowledge and Motivation

Adams and Sasse [4] investigated users' perceptions of password mechanisms to identify the human and organisational factors that can have an impact on the security and usability of password mechanisms. They found that many users' knowledge was shockingly inadequate, which leads to them constructing their own, often wildly inaccurate models of possible security threats and the importance of security. This, in turn, gives rise to a wide variety of user behaviour that is undesirable in terms of security. Studies 1 and 3 confirmed this state of affairs. It is therefore not surprising that 'user education' has appeared on top of the to-do list of many security departments. However, this will require more than pushing documents to users.

From the task perspective (see section 15.4), security is an enabling task, i.e. one that needs to be completed in order to be able to perform the main task. Authentication is an enabling task that needs to be completed to get to the resources required to do real work. It is easy to see that for many users today, security is something that gets in the way of real work, especially when mechanisms are difficult to use, and/or the need for security is not obvious. Users have to be motivated to make the additional effort that is required to use security mechanisms properly.

Small business or home users can be motivated to a certain extent by education about the risks of not bothering with security. This allows them to make an informed choice about their behaviour, based on an assessment of the risks to them personally and the effort required to reduce these risks. This is different for large organisations (see section 15.5). Users are generally less concerned about security of their organisation's computer systems, and often choose to follow existing policies only selectively or not at all.

In order to develop effective means of educating and motivating users with respect to password mechanisms, we tried to understand the mindset of users at which such measures will be targeted. In study 3, we identified seven issues that lead to undesirable password behaviour.

- Identity issues

  People who exhibit good password behaviour are often described as 'paranoid', 'pedantic' or 'the kind of person who doesn't trust anybody' — even by themselves. Some participants are proud of the fact that they do not understand security ('I'm not a nerd'), or do not comply with regulations ('I don't just follow orders').

- Social issues

  Sharing your password is considered by many users to be a sign of trust in their colleagues — you share your password with people you trust, and somebody refusing to share their password with you is effectively telling you that they do not trust you. The same goes for other password-related behaviour, such as locking your screen when you leave your computer for a few minutes — you are telling your colleagues next to you that you do not trust them.

- 'Nobody will target me'

  Most users think the data stored on their system is not important enough to become the target of a hacker or industrial spy. Hackers, for example, are assumed to target 'rich and famous' people or institutions. Some users accept that hackers might target somebody like them just to get into the overall system and try to get further from there, but clearly do not regard this as very likely, purely because of the number of users that could be targeted this way.

- 'They could not do much damage anyway'

  Most users do not think that somebody getting into their account could cause any serious harm to them or their organisation.

- Informal work procedures

  Current password mechanisms and regulations often clash with formal or informal work procedures (e.g. if you are ill and cannot come in, somebody in your group should be able to access your account and take care of your customers).

- Accountability

  Most users are aware that their behaviour does not fully comply with security regulations. However, they do not expect to be made accountable because they regard the regulations as 'unrealistic' and their behaviour as 'common practice'.

  In addition, they know that there is always a chance that a hacker will break into their system, however well they behave. They can always claim that it was not their misbehaviour that led to the break-in.

  Some users realise that they might be held accountable for past behaviour (e.g. writing down their password) if somebody gets into their system now.

- Double-binds

  If a computer system has strong security mechanisms, it is more likely to come under attack from hackers who want to prove themselves, and who will in the end find a way to get in. If it has weak security, inexperienced hackers, who try to break into many systems without targeting one specifically, might get in.

  Similarly, if you follow rules (e.g. lock your screen), people will think you have valuable data on your computer, and are more likely to try to break in. If you do not follow regulations, it is easier for somebody to break in.

An additional interesting result of study 3 was that there are users who show good behaviour even though they are of the opinion that this is not necessary. They follow regulations because they perceive it as necessary to maintain their professional reputation, or because they believe that any security failure involving their employer would ultimately reduce its standing in the business world. This insight provides a potential basis for further educational and motivational measures. A more drastic approach would be to link organisational security to users' personal data, by providing access to payroll, health records and personal e-mail — all users in study 3 stated such data was worthy of good security behaviour.

15.4 Goals and Tasks

Consideration of user's goals and tasks is the key element of a user-centred approach to design. Technology is not an end in itself, but should work in tandem with users to achieve an individual's or organisation's goals; these goals need to be identified to ensure that user/system interaction is designed to be effective, i.e. produces the required output. Goals are achieved through the completion of tasks; technology designers study tasks to ensure that user-system interaction is designed to be efficient, i.e. can be completed as quickly as possible and without waste of resources. An analysis of goals and tasks typically identifies:

- the goals (desired output);
- fundamental tasks (without which the desired output cannot be achieved);
- enabling tasks (which have to be completed in order to be able to carry out a fundamental task);
- performance criteria for each task (e.g. speed of completion, maximum number of errors);
- frequency with which each task is carried out;
- resources required by users and technology to carry out the task.

The insights gained through such analysis provide essential input into the design process. Fundamental tasks, for instance, are given priority in terms of visibility and feedback, and frequent tasks need to be particularly efficient in execution. Another fundamental principle of good design is to apportion tasks to users and technology in line with their strengths and weaknesses (e.g. limitations of human memory, see section 15.3.1).

15.4.1 Tasks and Passwords

Many of the problems that users have with security mechanisms can be explained in terms of a bad match between the mechanisms and users' goals and tasks. Users' behaviour is essentially goal-driven. If the benefits of an enabling task are not obvious to users, they will view it as something that gets in the way of completing the fundamental task, and find ways of cutting it out if possible. The studies by Whitten and Tygar [19] and Adams and Sasse [4] described this phenomenon for PGP encryption and password mechanisms, respectively. Whether or not it is possible to bypass an enabling task, the extra effort required will foster resentment in users, and encourage the perception that security is 'not sensible' because it interferes with real work. This, in turn, reduces user motivation (see section 15.3.2), which in the longer terms leads to an erosion of security culture (see section 15.5.2).

All our studies provided evidence of how badly password mechanisms are currently matched to users' capabilities and their tasks. This is because password mechanisms, and the policies that govern their use, are currently implemented as general mechanisms to protect access to systems, and without reference to the work that is being performed. Infrequently used passwords, for example, would be better served by a mechanism that does not require 100% accurate recall of strong memory items, i.e. accepts partly successful authentication, a combination of weak items, or relies on recognition (see sections 15.3.1.1 and 15.3.1.3). More tolerant mechanisms would also work better in conjunction with high-speed, high-pressure tasks, which have an increased likelihood of user slips.

Current mechanisms generally do not acknowledge the cost of authentication failure for fundamental tasks. Users and organisations can suffer significant losses as a result of not being able to access a system needed for a fundamental task because of authentication failure. If there is no contingency when legitimate users are unable to gain access, they are left to invent their own, such as borrowing a colleague's password — there was plenty of evidence for this in studies 1 and 3. There may be help desks or system administrators, but if a re-set takes 15 mins to complete, and an important customer wants a quote now, this is not a valid contingency.

Consequently, security must be designed as an integral part of the system that supports a particular work activity in order to be effective and efficient. Decisions about system and file access must be based on how tasks and workflow are organised in the real world. In modern organisations, many tasks are assigned to teams, and teamwork and collaboration are encouraged. If users are then given individual passwords, and unable to access each other's files even though they are needed for shared tasks, password disclose will become common. Finally, a task-centred design recognises that support resources, such as instructions and help, need to be available at the point where users need them. Instructions for constructing and memorising a strong password, for instance, should be available when a password needs to be chosen or changed — and the instructions, as well as labels on any tools, need to be compatible with users' vocabulary [19].

15.5 Context

An effective and efficient design of a user-technology interaction will be largely determined by the goals of that interaction and the tasks required to complete it. Beyer and Holtzblatt [20] identified a host of physical and social factors that need to be considered and accommodated to design an effective system. This section presents a number of points that are relevant to password security.

15.5.1 Physical Environment

The physical environment in which password mechanisms are used can influence user behaviour. Adams and Sasse [4] found that if the physical security environment has obvious flaws, users may feel that it is not worth bothering with passwords ('after all, anyone can get in here'). A strong physical security environment can lead to complacency, which needs to be counteracted by reminding users of risks facing the organisation (see section 15.5.2).

Presence of others is an important consideration:

- users may worry about how they appear to others (e.g. 'paranoid', see section 15.3.2);
- many users become nervous when they feel observed by others, which can have a negative impact on their ability to recall and enter passwords accurately — many people feel under pressure when there is a queue of other users waiting to use a cash dispenser, for instance, and feel embarrassed that others can see they have a problem, perhaps feeling embarrassed because others:

    — can observe their ineptitude;

    — may assume that they are trying to gain access to something to which they are not entitled.

15.5.2 Social and Organisational Environment — Security Culture

Study 3 provided in-depth data that showed that good password behaviour can lead to social repercussions (see section 15.3.2). Users who behave according to regulations are seen as 'paranoid' or 'pedantic', and anybody not willing to share their password with colleagues might be regarded as untrusting, and possibly even untrustworthy. All this points to an important area of further research. For effective security, organisations must develop a culture in which passwords are not only integrated into people's work (see section 15.4), but security is adopted as a shared concern by all employees. In many organisations, there is a wide gap between security policies and widespread insecure behaviour. Password rules that are unworkable in practice cannot be enforced and are thus not taken seriously; since security provides the rationale for password rules, it also ends up not being taken seriously. Highly paid key staff often feel that they are too busy to obey 'petty' password rules, and those in charge of security are often not in a position to enforce compliance from these staff. A sinister side-effect on security culture is that being able to flaunt security regulations becomes a badge of seniority.

The first step towards recovering security culture is to ensure that password mechanisms are not unworkable. Security design has to integrate all aspects of security, from the technical to the user interface and user training, with the organisation's work practices and overall culture. We believe that security policies, and the way in which they are presented and enforced, are a fundamental leverage point that makes it possible to move towards such an integration of security and overall organisational culture. However, it will take significant work in terms of user education and motivation to achieve a state where all members of the organisation accept their role in, and responsibility for, the security of an organisation. We have adapted the use of fear appeals [21] as a means of convincing users that good security behaviour also serves their own interest [22]. The main points of this approach are threefold.

- Impact of security on business

  Emphasise the importance of security for the organisation's business. Show how the organisation's reputation and business would be affected if it becomes known that employees engage in behaviour which, for instance, might endanger confidentiality of customer data. Most employees realise that lost business means jobs in danger. This gives the 'fear appeal' (and the associated punishment) a rational motivation that will raise users' acceptance of it.

- Punishment

  Appropriately punish behaviour, not its consequences. Make it clear that you cannot monitor all the employees all the time, but that you will make detailed enquiries about their past behaviour in a case of break-in through their account. This behaviour will definitely be punished, whether it led to the actual break-in or not.

- Security awareness

  Report security transgressions, rather than trying to keep them secret in an attempt not to lose face. Currently, there are few rewards for security-conscious behaviour; if regulations are to be taken seriously, failure to observe them must be dealt with, and seen to be dealt with. This is effectively learning by negative reinforcement, which can only be effective if security failures are made known to users.

15.6 Summary

15.6.1 Do Passwords Have a Future?

We have identified a considerable number of usability issues with password mechanisms. At the same time, our analysis has revealed that many problems are due to the way in which passwords are currently implemented.

- Single sign-on

  Organisations can fight password inflation by moving away from designing security on a per-system basis. SSO, a single userid, and password rules that are consistent across systems, can help achieve this. Also, security must be designed as an integral part of users' work, rather than get in the way.

- Reducing forced changes

  Judicious changes to password policies can make strong passwords manageable. Organisations can curb password inflation by reducing forced changes, and sanctioning use of the same password for several systems. In our view, this entails fewer risks than abandoning the cardinal principle of knowledge-based authentication and moving to wholesale writing down of passwords, as Schneier suggests [1]. Passwords that are written down are harder to protect than those that are memorised.

- Alternatives

  There are alternatives to the two-step password procedure, which requires unaided recall of a strong password, and these would be more suitable for infrequently used passwords.

- Password management

  There are techniques for designing and managing a certain number of strong passwords; these techniques need to be made available when users need them.

- User motivation

  User education will only work if users are motivated.

Nielsen [8] suggests that passwords are a usability nuisance that will be abolished by wide-scale introduction of biometric authentication. Biometric systems may be a good fit for some user/tasks/context configurations, but not all of them (see Chapter 14). We predict that knowledge-based authentication, in a more appropriate form than today, will be used in the foreseeable future.

15.6.2 Does Security Need Special Usability?

Whitten and Tygar [19] suggested that security needs a usability standard that is different from those applied to 'general consumer software'. We disagree with this view — after all, Whitten and Tygar themselves used standard HCI methods to uncover the usability problems in the user interface of PGP. In this chapter, we have outlined how existing HCI knowledge and tools, properly applied, would go a long way towards addressing usability issues in security.

If there are gaps in the HCI repertoire when it comes to designing usable security, they are in the area of:

- user motivation;
- a design method that integrates the different aspects that affect usability of security.

We have adapted the fear appeals from cognitive therapy to make security instructions more persuasive [23], and are currently running user trials to test the policies, tutorials and user interfaces that incorporate persuasion. To devise a design method for the whole socio-technical system that is security, we are currently adapting Reason's framework [3] for the design and maintenance of safety-critical systems [22]. Safety and security protect the interests of both individuals and organisations in the long term. However, they also compete for resources with fundamental tasks in the short term, and are likely to be sacrificed for short-term gain. This tendency can only be overcome by making security a visible and integral part of an organisation's long-term goals and its daily activities.

References

1 Schneier, B.: '*Secrets and Lies*', John Wiley and Sons (2000).

2 Poulsen, K.: '*Mitnick to lawmakers: People, phones and weakest links*', (March 2000) — http://www.politechbot.com/p-00969.html

3 Reason, J.: '*Human Error*', Cambridge University Press, Cambridge, UK (1990).

4 Adams, A. and Sasse, M. A.: '*Users are not the enemy*', Communications of the ACM, **42**(12) (December 1999).

5 Brostoff, S. and Sasse, M. A.: '*Are Passfaces more usable than passwords? A field trial investigation*', in McDonald, S. et al (Eds): '*People and Computers XIV — Usability or Else*', Proceedings of HCI, Sunderland, UK, pp 405-424, Springer (September 2000).

6 FIPS: '*Password Usage*', Federal Information Processing Standards publication (May 1985).

7 Adams, A., Sasse, M. A. and Lunt, P.: '*Making passwords secure and usable*', in Thimbleby H et al (Eds): '*People and Computers XII*', Proceedings of HCI'97, Bristol, Springer (August 1997).

8 Nielsen, J.: '*Security and Human Factors*', Alertbox (November 2000) — http://www.useit.com/alertbox/20001126.html

9 Haskett, J. A.: '*Pass-algorithms: a user validation scheme based on knowledge of secret algorithms*', Communications of the ACM, **27**(8), pp 777-781 (1984).

10 Zviran, M. and Haga, W. J.: '*A comparison of password techniques for multilevel authentication mechanisms*', The Computer Journal, **36**(3), pp 227-237 (1993).

11 Zviran, M. and Haga, W. J.: '*Cognitive passwords: the key to easy access control*', Computers and Security, **9**(8), pp 723-736 (1990).

12 Ellison, C., Hall, C., Milbert, R. and Schneier, B.: '*Protecting secret keys with personal entropy*', — http://www.counterpane.com/personal-entropy.pdf

13 Spector, Y. and Ginzberg, J.: '*Pass sentence — a new approach to computer code*', Computers and Security, **13**(2), pp 145—160 (1994).

14 Passlogix® Inc — http://www.v-go.com/

15 Dhamija, R., Perrig, A. and Deja, V.: '*A User Study — Using Images for Authentication*', Proceedings of the 9th USENIX Security Symposium, Denver, Colorado (2000).

16 Passfaces™ — http://www.idarts.com/

17 Valentine, T.: '*An evaluation of the Passface™ personal authentication system*', (Technical Report) Goldsmiths College, University of London (1998).

18 Valentine, T.: '*Memory for Passfaces™ after a long delay*', (Technical Report) Goldsmiths College, University of London (1999).

19 Whitten, A. and Tygar, J. D.: '*Why Johnny can't encrypt: A usability evaluation of PGP 5.0*', Proceedings of the 8th USENIX Security Composium, Washington (August 1999).

20 Beyer, H. and Holtzblatt, K.: '*Contextual design*', Morgan Kauffmann (1997).

21 Rogers, R. W.: '*A protection motivation theory of fear appeals and attitude change*', The Journal of Psychology, **91**, pp 93-114 (1975).

22 Brostoff, S. and Sasse, M. A.: '*Safe and sound: a safety-critical design approach to security*', to be presented at the 10th ACM/SIGSAC New Security Paradigms Workshop, Cloudcroft, New Mexico (September 2001).

23 Weirich, D. and Sasse, M. A.: '*Pretty good persuasion: a first step towards effective password security for the real world*', to be presented at the 10th ACM/SIGSAC New Security Paradigms Workshop, Cloudcroft, New Mexico (September 2001).

16

SECURITY MANAGEMENT STANDARD — ISO 17799/ BS 7799

M J Kenning

16.1 Introduction

Security is more than using the right technology. In the words of cryptographer Bruce Schneier: 'If you think technology can solve your security problems, then you don't understand the problems and you don't understand the technology' [1]. Security is as much about people, and the way they use the technology. The information security management standard, BS 7799 [2, 3], addresses this very issue.

BS 7799 was developed in the early 1990s as a result of demand from industry, government and commerce for a common information security framework. Organisations felt that they needed to assure those with whom they do business that they operate to a common minimum security standard. They also needed to be able to provide others with assurances about their own security.

A group of companies, including BOC, BT, Marks and Spencer, Midland Bank, Nationwide Building Society, Shell and Unilever, co-operated in the development of the Code of Practice for Information Security Management — BS 7799 Part 1 Code of Practice. The Specification for Information Security Management Systems — BS 7799 Part 2 — was published in February 1998 [2, 3]. Part 1 of the standard was published as the international standard ISO/IEC 17799 Part 1 code of practice for information security management in December 2000 [4].

In the UK the scheme for accredited certification of an organisation's information security management system (ISMS) to the requirements of BS 7799, is known as 'c:cure'. The scheme, commissioned by the DTI in 1998 and managed by BSI-DISC, requires participating certification bodies to be accredited by recognised national accreditation bodies for this activity [5]. c:cure also requires that the auditors used by a certification body for assessment of organisations against the scheme criteria are registered specifically for this activity with a recognised auditor registration scheme, such as that offered by the International Register of Certified Auditors.

A certificate issued under the scheme is valid for three years, subject to satisfactory maintenance of the system, which will be checked during surveillance visits at least annually. Thereafter, certificates will typically be renewed for a further three years.

In short, conformance to BS 7799 is a matter of putting in place appropriate security controls in the first instance, coupled with ongoing monitoring and improvements to ensure that the controls remain effective and appropriate. The decision as to what is appropriate depends upon understanding the risks and costs involved. Understanding the risk means knowing what the assets are, what the possible threats to those assets are, and the likelihood and possible impact of a security breach on the business.

16.2 The Standard

BS 7799 is designed to assure the confidentiality, integrity and availability of information assets. This is achieved through security controls implemented and maintained within the organisation. The key areas identified by BS 7799 for the implementation of an information security management system are:

- an information security policy;
- allocation of information security responsibilities within the organisation;
- asset classification and control;
- personnel security, responsibilities and training;
- physical and environmental security;
- communications and operational systems security;
- access controls;
- systems development and maintenance;
- business continuity;
- periodic compliance reviews.

The information security management system includes the following:

- scope statement;
- security policy document;
- asset list;

- risk assessment;
- statement of applicability;
- business continuity plan.

The information security management system will also include processes which continually monitor the effectiveness of security protection for the information and associated systems.

BS 7799 is not a once and for all process — there is a requirement for continual monitoring and ongoing improvement. Nor is it a substitute for evaluation criteria such as the Common Criteria [6] or the ITSEC scheme [7]. There could, however, be an implicit level of trust placed in a product or system created or run by an organisation which is BS 7799/ISO 17799 certified.

16.3 Case Study — Certification of the SETT

Following the successful certification of BT Security's policy set in 1999, it was decided to seek certification for a unit within BT. The benefit to be gained, in addition to the intrinsic benefits of the process improvements involved, was anticipated to be the insights and experience which could then be fed out to other parts of BT wishing to go down that route. The plan was successful and the Security Evaluation and Test Team (SETT) became the first unit in BT to achieve certification to BS 7799.

The Security Evaluation and Testing Team is a team of five people whose main area of responsibility is to work with BT Security in the application of the BT Security Evaluation and Certification Scheme (BTSECS). An ambitious plan was drawn up (Table 16.1) and the process adopted is illustrated in Fig 16.1.

Table 16.1 SETT certification project plan.

1 November 1999	**Workpackage start** Understand BS 7799 requirements
13 December 1999	**Consultancy input** Develop 1st draft of 7799 document set
1 February 2000	**Gap analysis by external auditors** Review and issue 1.0 of 7799 document set
2-3 March 2000	**Audit stage 1** Corrective actions Review and issue 1.0 of 7799 document set
29-30 March 2000	**Audit stage 2**

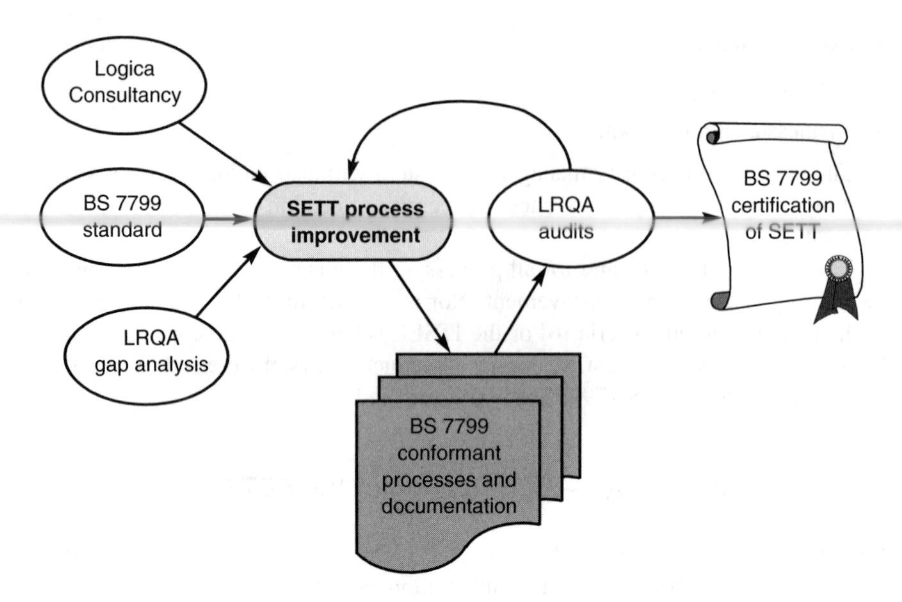

Fig 16.1 SETT certification process.

To achieve certification, and to carry out the ongoing management of the ISMS it was decided to create a Security Forum. The principal duties of the Security Forum are measuring and continually improving compliance with the ISMS (see Fig 16.2). These duties include:

- setting the ISMS scope;
- reviewing and approving all documentation associated with the ISMS;
- carrying out risk assessment and recording all changes;
- adopting security controls which reduce the security risk, consistent with the commercial imperatives of the team;
- appointing a security manager;
- conducting compliance checks against the ISMS;
- implementing and monitoring corrective actions arising from compliance checks;
- reviewing security incidents and producing a log of incident reports;
- reviewing security news.

A major challenge facing the team was that, as part of a large company, the team had little control over large areas of policy and process relating to security which were set company wide. For example, BS 7799 has specifications relating to the recruitment of personnel, but such matters are dealt with by the company human resources department. Similarly accommodation and IT infrastructure, such as file servers, local and wide area networking provision, are all handled at company level.

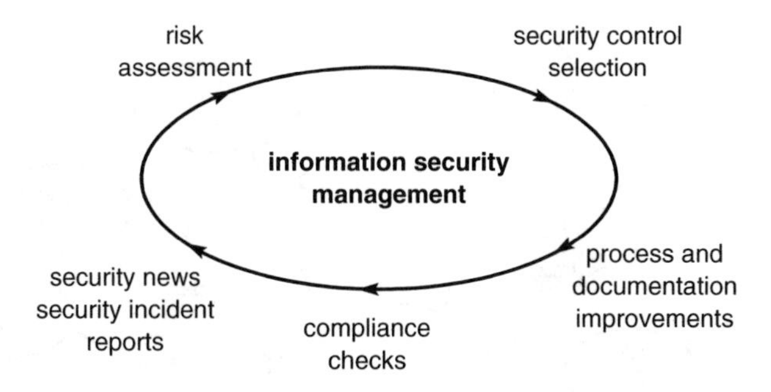

Fig 16.2 Security Forum process.

The approach adopted by the team was to identify all such interfaces between the team and external environment (including the rest of the company) and examine the service level agreements for each interface (see Fig 16.3). If the level of service was deemed to be adequate, no further action was taken. If not, a decision was taken to supplement the service in an appropriate way and a suitable control implemented.

BS 7799 is not a quality standard as such but assumes relevant actions are done in a quality way. The philosophy 'Say what you do and do what you say' underpins the audit process. It is necessary to be able to show that information security management decisions have been made and followed up. This implies quality processes for recording and monitoring work, but no specific processes or recording mechanisms are mandated.

A major difference between BS 7799 and ISO 9001, say, in the experience of the team, was that the standard has built-in best practice security requirements. This means that there are assumed controls to be either adhered to or explained away. The standard can be thought of as a quality standard with a security attitude.

16.4 Technical Implications

BS 7799 Part 2 includes the following requirement in Clause 3.2c [3]:

> 'Appropriate control objectives and controls shall be selected from clause 4 for implementation by the organisation, and the selection shall be justified.'

Clause 4 of the standard sets out the detailed security controls corresponding to the key areas listed in section 16.2 above. In addition to procedural and personnel security, these controls imply the use of technical security measures, including:

- virus detection and prevention;
- protection of the privacy of personal data;
- use of encryption, for example to protect data on laptops;

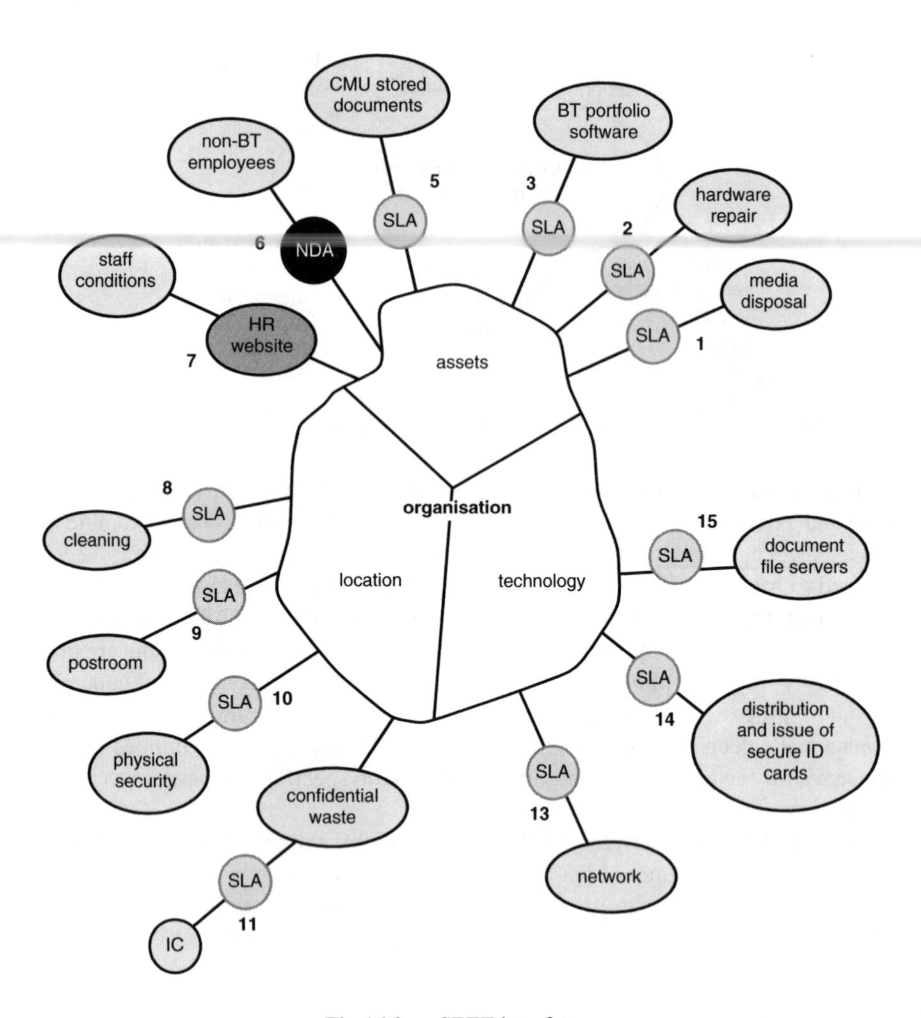

Fig 16.3 SETT interfaces.

- network security, including access controls, user segregation and routing controls.

Achieving BS 7799 certification can therefore involve rolling out appropriate technical controls to protect information on systems used by the organisation, and security upgrade to the organisation LAN.

In the case of the SETT, the team was able to demonstrate that it benefited from the inherent security of BT's intranet. This includes access control to the LAN, site-wide FDDI ring, and switched Ethernet within buildings. The use of fibre distribution and switching technology raises the level of confidence that packets cannot be intercepted or interfered with (see Fig 16.4).

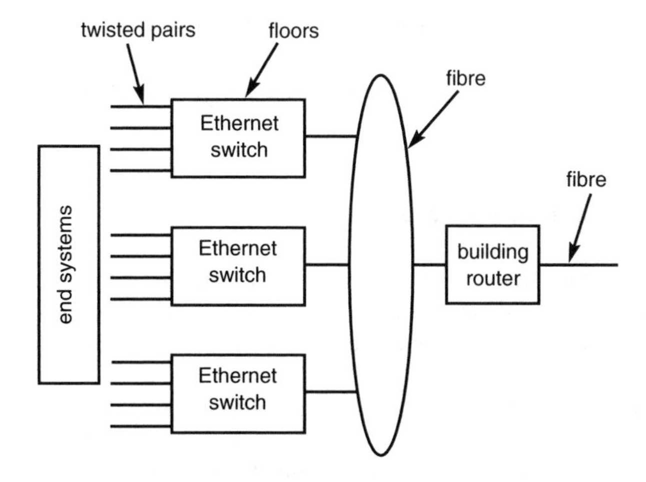

Fig 16.4 Generic switched Ethernet.

16.5 Commercial Advantage

A commercial organisation looks at a proposed innovation from two standpoints — does it drive down costs, does it drive up revenues. There are several reasons why a commercial organisation would be considering BS 7799 certification:

- its customers are asking for BS 7799;
- its customers include government, banks or financial institutions;
- it is developing eBusiness applications;
- it wants to take information security seriously.

16.5.1 Cost of Security

While it is true that security in general, and certification to BS 7799 in particular, imply a cost, it is also true that inadequate security can also be very costly.

The Information Security Breaches Survey 2000 Technical Report [8] states that 60% of the organisations surveyed had experienced a security breach in the 2 years since the previous survey. This had increased dramatically from the 1998 figure of 40%. The same report states that: '... very few organisations were able (or prepared) to report the business implications of the security breaches they had suffered — but those that were, indicated that the cost of a single breach could be in excess of £100,000'. A previous survey had put the average cost of a security breach at £16,000 [9].

The 'Love Bug' virus of 2001 is reckoned to have caused $2.6 billion worth of damage world-wide.

Avoidance of just one serious security breach per year would probably cover the cost of BS 7799 conformance.

16.5.2 Market for Security

The potential future market for security services is considerable. According to GartnerGroup, world-wide eCommerce will grow to approximately $7.3 trillion in 2004 [10]. This growth in eCommerce will entrain massive demand for security consultancy and software.

Dresdener Kleinwort Benson, based on information gathered from various industry sources, estimates that the Internet security software market will reach $13.3bn by 2004, equating to a 24% compound annual growth rate. Further they estimate that the market for global eSecurity will grow from under $1bn in 1999 to over $10.8bn over four years — an annualised rate of 68% [11].

16.5.3 Government Attitude to BS 7799

The UK Government is committed to BS 7799, both internally and as a requirement to be placed on suppliers of IT services to Government. The Cabinet Office has instructed each government department to put plans in place to show how and when they will implement BS 7799 [12].

The document entitled 'A Commercial and Policy Framework for the Third Party Delivery of Government Services' is part of a set of documents published in conjunction with the Modernising Government white paper. That document [13] states that:

> 'The holder of a licence will be certified as compliant with BS 7799 in terms of security policy, personnel security, physical and environmental security, computer and network management, system access control, business continuity planning.'

16.5.4 The Data Protection Act 1998

The Data Protection Act 1998 lays out a set of eight principles and effective information security is implicit in all of them [14]. However, it is Principle 7, relating to the prevention of unauthorised or unlawful processing, and of accidental loss or damage to data, which deals most directly with the need for confidentiality, integrity and availability of data. All organisations have to comply with the eight principles.

Putting in place a BS 7799-conformant ISMS can be a cost-effective way of inculcating an information security culture throughout the organisation. At the very least, an organisation which has an information security management system will be in a better position to demonstrate compliance with the Act.

16.6 Summary

Gaining BS 7799 Certification for the Security Evaluation and Test Team was beneficial because it has resulted in process improvements in the way the team works and also provides an external validation for the quality and effectiveness of the team's professional expertise in information security management. This is seen as a useful selling point for the services of the team in the commercially independent climate in which the company increasingly intends to operate in the future.

BTexact Technologies is BT's technology business and therefore information security is crucial both in defence of the information assets of the business and reassuring customers that their information is in safe hands. BS 7799 certification is an independent mark of quality in the external market-place.

References

1 Schneier, B.: '*Secrets and Lies*', John Wiley and Sons, New York (2000).

2 BS 7799: '*The Standard for Information Security Management*', (April 2001) — http://www.c-cure.org/fbs7799.htm

3 BS 7799-2:1999: '*Information Security Management — Part 2: Specification for Information Security Management Systems*', (April 2001).

4 ISO 17799 (April 2001) — http://www.bsi-global.com/Information+Security/04_Standards_infosec/index.xhtml

5 Accredited Certification for BS 7799, UKAS (April 2001) — http://www.ukas.com/new_docs/technical-bs7799.htm

6 ISO/IEC 15408-x: '*Common Criteria — Evaluation Criteria for IT Security*', — http://csrc.nist.gov/cc/

7 '*Information Technology Security Evaluation Criteria (ITSEC)*', Version 1.2 (April 2001) — http://www.itsec.gov.uk/

8 Information Security Breaches Survey 2000 Technical Report (April 2001) — http://www.trustwise.com/repository/PDF/dtibro.pdf

9 DTI: '*The Business Manager's Guide to Information Security*', (April 2001) — http://www.dti.gov.uk/cii/datasecurity/businessmanagersguide/why_is_info_security_important_to_me.shtml

10 '*The Internet Security Investor Handbook*', Lehman Brothers (March 2001).

11 '*Spreading the Web of Trust — eSecurity*', Dresdener Kleinwort Benson (October 2000).

12 '*BS 7799 and the Cabinet Office*' (April 2001) — http://www.bcs.org.uk/review/2001/html/p185.htm

13 UK Government: '*Channels for Electronic Service Delivery: an Outline of the Policy*', (April 2001) — http://www.iagchampions.gov.uk/moderngov/cppolicy10.htm

14 UK Government: '*BS 7799 And The Data Protection Act 1998*', (April 2001) — http://www.dti.gov.uk/cii/datasecurity/1998dataprotectionact/7799_data_protection.shtm

ACRONYMNS

24 × 7	24 hours a day, 7 days a week
3GPP	Third Generation Partnership Project
ADSL	asymmetric digital subscriber line
AES	advanced encryption standard
AFR	automatic face recognition
AH	authentication header
AKA	authentication and key agreement
AMF	authentication management field
AN	access network
ANX	Automotive Network Exchange
API	application processing interface
ASP	application service provider
AuC	authentication centre
AUTN	authentication token
AUTS	re-synchronisation token
AV	authentication vector
BITS	bump in the stack
BITW	bump in the wire
CA	certification authority
CAMEL	customised applications for mobile network enhanced logic
CAP	CAMEL application part
CCK	common cipher key
CDb	certificate database
CERT	computer emergency response team
CESG	Communications-Electronics Security Group

CFB	customer-focused billing
CHTML	compact HTML
CK	cipher key
CN	core network
CNI	critical national infrastructure
CORBA	Common Object Request Broker Architecture
CP	certificate policy
CPS	certificate practices statement
CRL	certificate revocation list
CRS	certificate request syntax
CS	circuit switched
CSCF	call state control function
CSR	certificate signing request
CSS	customer service system
CTL	certificate trust list
CUG	closed user group
Cx	CSCF/HSS interface
DAS	digital archiving server
DCK	derived cipher key
DCOM	distributed component object model
DDOS	distributed denial of service
DECT	digital enhanced cordless telecommunications
DES	Data Encryption Standard
DMO	direct mode operation
DNS	domain name system
DPD	delegated path discovery
DPV	delegated path validation
DSV	dynamic signature verification
DVLA	Driver and Vehicle Licensing Agency (UK)
ebXML	eBusiness XML
ECA	Electronic Communications Act (UK)
EE	end entity

EESSI	European Electronic Signature Standardisation Initiative
EJB	Enterprise Java Bean
EMW	enterprise middleware
ESIGN	Electronic Signatures in Global and National Commerce Act (USA)
ESP	encapsulating security payload
ETSI	European Telecommunications Standards Institute
FDDI	fibre distributed data interface
FTP	file transfer protocol
GCK	group cipher key
GEK	group encryption key
GGSN	gateway support node
GPRS	general packet radio service
GSM	global system for mobile communications
GSSI	group short subscriber identity
GUI	graphical user interface
HCI	human-computer interaction
HE	home environment
HLR	home location register
HSS	home subscriber server
HTML	hypertext markup language
HTTP	hypertext transfer protocol
IAAC	Information Assurance Advisory Council
IAP	information assurance project
IBIA	International Biometric Industry Association
IDPKC	identifier-based public key cryptosystem
IDS	intrusion detection system
IETF	Internet Engineering Task Force
IK	integrity key
IKE	Internet key exchange
IM	IP multimedia
IMS	Internet multimedia subsystem
IMSI	international mobile subscriber identity

IP	Internet protocol
ISMS	information security management system
ISO	International Standards Organisation
ISP	Internet service provider
ITSI	individual terminal subscriber identity
JAAS	Java Authentication and Authorisation Service
JSP	Java Server Page
KEK	key encryption key
L3	layer 3
LA	location area
LAN	local area network
LDAP	lightweight directory access protocol
LoB	line of business
LRA	local registration authority
MAC	message authentication code
ME	mobile equipment
MAP	mobile application part
MExE	mobile station execution environment
MGCK	modified GCK
MoU	memorandum of understanding
MS	mobile station
MSC	mobile services switching centre
MSP	managed service provider
NAT	network address translation
NCIS	National Criminal Intelligence Service
NCSA	National Center for Supercomputing Applications
NISCC	National Infrastructure Security Co-ordination Centre
NNTP	network news transport protocol
OASIS	Organisation for the Advancement of Structural Information Systems
OCSP	online certificate status protocol
OOB	out-of-band
OSA	open service access

OSI	open systems interconnection
OSPF	open shortest path first
OSS	operational support systems
OTAR	over-the-air rekeying
PCK	primary correlation key
PCT	private communication layer protocol
PDA	personal digital assistant
PIN	personal identification number
PKC	public key cryptography
PKI	public key infrastructure
PKIXCMP	X.509 PKI certificate management protocol
POT	pyramid of trust
PS	packet switched
PSTN	public switched telephone network
RA	registration authority
RAND	random challenge
RAS	Remote Access Service
RES	(expected) user response
RNC	radio network controller
S2ML	security services markup language
SA	security association
SAML	security assertion markup language
SAX	simple API for XML
SCK	static cipher key
SCVP	simple certificate validation protocol
SDS	short data service
SFPG	Security and Fraud Prevention Group (TETRA)
SGSN	serving GPRS support node
SIM	subscriber identity module
SIP	session initiation protocol
SME	small/medium enterprise
SMTP	simple mail transfer protocol

SN	serving network
SOAP	simple object access protocol
SPKI	secure PKI
SQN	sequence number
SQN_{HE}	sequence number counter maintained in the HLR/AuC
SQN_{MS}	sequence number counter maintained in the USIM
SSL	secure sockets layer
SSO	single sign-on
STK	SIM Tool-kit
STLP	secure transport layer protocol
SVG	scalable vector graphics
TAA	TETRA authentication algorithm
TCP	transmission control protocol
TEA	TETRA encryption algorithm
TEI	TETRA equipment identity
TEK	traffic encryption key
TETRA	terrestrial trunked radio standard
TLS	transport layer security
TMSI	temporary mobile subscriber identity
UDDI	Universal Description, Discovery and Integration Specification
UE	user equipment
UEA	UMTS encryption algorithm
UETA	Uniform Electronic Transaction Act
UIA	UMTS integrity algorithm
UICC	UMTS IC card
UMTS	universal mobile telecommunications system
URI	universal resource identifier
URL	uniform resource locator
USIM	UMTS subscriber identity module
VA	validation authority
VB	validation broker
VIP	virtual IP

VLR	visitor location register
VPN	virtual private nerwork
W3C	World Wide Web Consortium
WAN	wide area network
WAP	wireless application protocol
WIM	WAP identity module
WML	wireless markup language
WPKI	WAP PKI
WSDL	Web services description language
WTLS	wireless transport layer security
WWW	World Wide Web
XACML	extensible access control markup language
XAML	transaction authority markup language
XHTML	extensible hypertext markup language
X-KISS	XML key retrieval service standard
XKMS	XML Key Management Specification
X-KRSS	XML key registration service standard
XML	extensible markup language
XRES	expected response
XSL	XML stylesheet language
XSLT	XSL transformation
X-TASS	XML Trust Assertion Service Specification (VeriSign)

INDEX